Integrable
Many-Particle
Systems

Integrable Many-Particle Systems

Vladimir I Inozemtsev

Joint Institute for Nuclear Research, Russia

World Scientific

NEW JERSEY · LONDON · SINGAPORE · BEIJING · SHANGHAI · HONG KONG · TAIPEI · CHENNAI · TOKYO

Published by

World Scientific Publishing Europe Ltd.

57 Shelton Street, Covent Garden, London WC2H 9HE

Head office: 5 Toh Tuck Link, Singapore 596224

USA office: 27 Warren Street, Suite 401-402, Hackensack, NJ 07601

British Library Cataloguing-in-Publication Data
A catalogue record for this book is available from the British Library.

INTEGRABLE MANY-PARTICLE SYSTEMS

ISBN 978-1-80061-381-2 (hardcover)
ISBN 978-1-80061-382-9 (ebook for institutions)
ISBN 978-1-80061-383-6 (ebook for individuals)

For any available supplementary material, please visit
https://www.worldscientific.com/worldscibooks/10.1142/Q0407#t=suppl

Typeset by Stallion Press
Email: enquiries@stallionpress.com

Printed in Singapore

Preface

It is of common knowledge that the problem of three or more particles interacting with forces described by two-body potential is not solved analytically (except in the trivial case of harmonic oscillators). Being true in two or more dimensions, it is NOT true in one dimension: there are a lot of examples when, for special interacting potentials, problems of many-body systems in one dimension are integrable in the Liouville sense (i.e. exactly solvable). Moreover, in one dimension there are exactly solvable quantum three- and many-body systems and, even more, those allow the property of quantum integrability which cannot be solved (since there is no analog of quantum Liouville theorem), but the spectral problem might be reduced drastically.

The first example of such a system was found by C. Jacobi in the last half of 19th century. He noted that three-particle systems interacting with the two-body potential depending on the inverse square of the distance between particles are exactly solvable since, after removing the motion of the center of mass (which is trivial) the remaining dynamical variables are separated in the polar system of co-ordinates. Being published in Latin, the Jacobi memorandum was completely forgotten and only in 1969 his result was re-opened by Italian physicist Francesco Calogero (also for three particles, but in the quantum case). Later, this result was extended (purely empirically) for arbitrary number of particles with equal mass and the same inverse-square interaction. In 1971, Bill Sutherland found, also completely empirically, that the problem of arbitrary number of particles of equal masses interacting by inverse-square trigonometric two-body potential, the full spectrum of quantum problem and algorithm for calculating wave functions (later it was mentioned that the similar functions were first described by the mathematician Jack in 1969, so now they are often named as Jack polynomials). The Sutherland problem only has a discrete spectrum — all particles are bounded. The problem with inverse-square hyperbolic two-body potential with scattering states only turned out much more complicated from the mathematical point of view and cannot be solved in elementary functions; nevertheless, it was successfully attacked by Dutch mathematicians G. Heckman and H. Opdam in the middle of the 1980s. Soon, in 1990, the analytical solution of hyperbolic quantum Sutherland problem was found by mathematicians O. Chalykh and A. Veselov in terms of elementary functions (exponents and hyperbolic cotangents) for special values of coupling constants in the form of double differential recurrence relations.

As for a classical system of that kind, great progress was made in the beginning of 1970s with the use of Lax relations. The famous Calogero Ansatz for the Lax pairs (resulting in elegant functional equation) allows him to find a complete set of integrals of motion for arbitrary number of particles of equal masses for not only inverse square, inverse-square hyperbolic and trigonometric potentials, but the most general potential given by the Weierstrass elliptic $\wp$ function. Independently these results were obtained by J. Moser, so the physicists and mathematicians use the name Calogero–Sutherland–Moser (CSM) systems. At the end of 1970s, M. Olshanetsky and A. Perelomov noted that the differences of coordinates of particles in such systems are connected with the systems of roots of semi-simple Lie algebras, and found a way to solve classical equations of motion in rational and trigonometric cases.

A number of cases, limits and generalizations lead to enormous growth of the papers devoted to the CSM systems. The author's goal is not of course to review all of them. There are excellent reviews by M. Olshanetsky and A. Perelomov (*Physical Reports* 71, 314 (1981) and *Physical Reports* 94, 313 (1983)) describing quite well the situation till 1983. I would like to describe mainly my own results, paying attention to most important other achievements. I would like to impress the reader by the beauty of mathematical results concerning the models of many light particles interacting with particles of infinite mass, models of interacting quasiparticles and point out many unsolved problems in this branch of mathematical physics.

Contents

Chapter 2. Quantum Systems　　　　　73

Chapter 3. Integrable Systems of Quasi-particles　　　123

Chapter 4. Integrable Systems of Particles with Spin　221

Chapter 1

Classical Systems Disconnected with Lie Algebras

1.1. Integrable Models of the Motion of Two Particles Interacting with Particle of Infinite Mass

Classical integrable one-dimensional systems have been a subject of comprehensive study (see review paper of Olshanetsky and Perelomov). There are also known integrable problems on the motion of particles of equal mass with one particle of infinite mass creating an external field defined by some potential (in the following, we shall name these models as systems of particles in the external field): The models are defined by the Hamiltonian

$$H = \sum_{i=1}^{n} \left(\frac{1}{2} p_i^2 + W(x_i) \right) + \sum_{i>j}^{n} V(x_i - x_j).$$ (1.1)

The previous examples include the Olshanetsky and Perelomov (1976) system:

$$V(x) = a/x^2, \quad W(x) = bx^2$$ (1.2)

and the Adler (1977) system

$$V(x) = \frac{a}{\sinh^2 \frac{1}{2}x}, \quad W(x) = b \exp(x).$$ (1.3)

Do other integrable systems exist with the Hamiltonian (1.1)? Here we present positive answer to this question for $n = 2$, when for integrability only one function K of variables p_1, p_2, x_1, x_2 is required to be found satisfying the equation

$$\{K, H\}_P = 0 \qquad (1.4)$$

($\{., .\}_P$ are the Poisson brackets).

Consider a Hamiltonian of general form

$$H = \frac{1}{2}(p_1^2 + p_2^2) + V(x_1, x_2)$$

and look for the function K as a polynomial in the momenta p_1, p_2:

$$K = \sum_{i=0}^{l} \sum_{m=0}^{2i} A_{im}(x_1, x_2) p_1^m p_2^{2i-m}, \qquad (1.5)$$

or

$$K = \sum_{i=0}^{l} \sum_{m=0}^{2i+1} A_{lm}(x_1, x_2) p_1^m p_2^{2i-m+1}. \qquad (1.6)$$

At $l = 1$ the systems with constants of motion (1.5) and (1.6) were studied by Whittaker (1927) and Holt (1982).

Due to (1.4), the functions $V(x_1, x_2)$, $\{A_{im}(x_1, x_2)\}$ should obey a system of partial differential equations of first order (generally, nonlinear).

In particular, for the functions (1.6), this system contains $l^2 + 3l + 2$ equations

$$\frac{\partial A_{l,2l}}{\partial x_l} = 0, \quad \frac{\partial A_{l0}}{\partial x_2} = 0,$$

$$\frac{\partial A_{lm}}{\partial x_1} + \frac{\partial A_{l,m+1}}{\partial x_2} = 0 \quad (m = 0, \ldots, 2l - 1),$$

$$\frac{\partial A_{s,2s}}{\partial x_1} = (2s + 2) A_{s+1,2s+2} \frac{\partial V}{\partial x_1} + A_{s+1,2s+1} \frac{\partial V}{\partial x_2},$$

$$\frac{\partial A_{s,0}}{\partial x_2} = A_{s+1,1} \frac{\partial V}{\partial x_1} + (2s + 2) A_{s+1,0} \frac{\partial V}{\partial x_2},$$

$$\frac{\partial A_{sm}}{\partial x_1} + \frac{\partial A_{s,m+1}}{\partial x_2}$$

$$= (m+2)\frac{\partial V}{\partial x_1}A_{s+1,m+2} + (2s+1-m)A_{s+1,m+1}\frac{\partial V}{\partial x_2},$$

$$(s = 1, \ldots, l-1; \quad m = 0, \ldots, 2s-1),$$

$$\frac{\partial A_{00}}{\partial x_1} = 2\frac{\partial V}{\partial x_1}A_{12} + A_{11}\frac{\partial V}{\partial x_2}; \quad \frac{\partial A_{00}}{\partial x_2} = A_{11}\frac{\partial V}{\partial x_1} + 2A_{10}\frac{\partial V}{\partial x_2}.$$

$$(1.7)$$

The functions A_{lm} are polynomials in x_1 and x_2 of degree $2l$ and $(2l+1)$ for (1.5) and (1.6), respectively; the compatibility condition of equations for $A_{l-1,m}$ is a linear equation of order $2l$ or $(2l+1)$ for $V(x_1, x_2)$, coefficients of which are determined by $\{A_{lm}\}$. Upon finding its solution dependent on $2l$ (or $2l+1$) arbitrary functions of one variable, equations for the other coefficients A_{lm} in (1.5) and (1.6) lead either to a system of functional equations or to one functional equation for (1.5), $l = 2$ or (1.6), $l = 1$.

Our aim is to find integrable systems with a Hamiltonian of the type (1.1) at $n = 2$:

$$V(x_1, x_2) = W_1(x_1) + W_2(x_2) + V_{12}(x_1 - x_2). \tag{1.8}$$

It can be shown that for (2.6) at $l = 1$ the above-mentioned functional equation has no non-trivial solutions if we take $V(x_1, x_2)$ to be the functions of the type (1.8). In the case of (1.5), $l = 2$ the structure of $V(x_1, x_2)$ like (1.8) may appear, if we put

$$A_{20} = A_{21} = A_{23} = A_{24} = 0, \quad A_{22} = \frac{1}{2}.$$

Then the remaining Equations (1.7) assume the form

$$\frac{\partial A_{12}}{\partial x_1} = \frac{\partial A_{10}}{\partial x_2} = 0, \quad \frac{\partial A_{12}}{\partial x_2} + \frac{\partial A_{11}}{\partial x_1} = \frac{\partial V}{\partial x_2},$$

$$\frac{\partial A_{10}}{\partial x_1} + \frac{\partial A_{11}}{\partial x_2} = \frac{\partial V}{\partial x_1}, \tag{1.9}$$

$$\frac{\partial A_{00}}{\partial x_1} = 2\frac{\partial V}{\partial x_1}A_{12} + A_{11}\frac{\partial V}{\partial x_2}, \quad \frac{\partial A_{00}}{\partial x_2} = \frac{\partial V}{\partial x_1} + 2\frac{\partial V}{\partial x_2}A_{10}. \tag{1.10}$$

Equations (1.9) can be easily solved,

$$A_{12}(x_1, x_2) = W_2(x_2), A_{10}(x_1, x_2) = W_1(x_1),$$
$$A_{11}(x_1, x_2) = \tilde{V}_{12}(x_1 + x_2) - V_{12}(x_1 - x_2),$$
$$V(x_1, x_2) = \tilde{V}_{12}(x_1 + x_2) + V_{12}(x_1 - x_2) + W_1(x_1)$$
$$+ W_2(x_2). \tag{1.11}$$

Substituting (1.11) into (1.10), we arrive at the functional equation for $W_1, W_2, V_{12}, \tilde{V}_{12}$:

$$[\tilde{V}_{12}(x_1 + x_2) - V_{12}(x_1 - x_2)][W_2''(x_2) - W_1''(x_1)]$$
$$+ 2[\tilde{V}_{12}''(x_1 + x_2) - V_{12}''(x_1 - x_2)]$$
$$\times [W_2(x_2) - W_1(x_1)] + 3\tilde{V}_{12}'(x_1 + x_2)[W_2'(x_2) - W_1'(x_1)]$$
$$+ 3V_{12}'(x_1 - x_2)[W_2'(x_2) + W_1'(x_1)] = 0. \tag{1.12}$$

We do not know the general solution to this equation. Some particular solutions for which all four functions $V_{12}, \tilde{V}_{12}, W_1, W_2$ are different from zero correspond to systems studied by Olshanetsky and Perelomov at $n = 2$:

$$V_{12}(x) = \tilde{V}_{12}(x) = g^2 \wp(ax),$$
$$W_1(x) = W_2(x) = g_1^2 \wp(ax) + g_2^2 \wp(2ax), \tag{1.13}$$

where $\wp$ is the Weierstrass elliptic function with arbitrary periods $2\omega_1$, $2\omega_2$ and either $g_1^2 - 2g^2 + \sqrt{2}gg_2 = 0$, $g_1 \neq 0$, or $g_1 = 0, g, g_2$ are arbitrary.

Since Equation (1.12) is *linear* in W_1, W_2, more general solution to (1.12) takes place:

$$V_{12}(x) = \tilde{V}_{12}(x) = g^2 \wp(ax),$$
$$W_1(x) = W_2(x) = h_1 \wp(ax) + h_2 \wp(a(x + \omega_1))$$
$$+ h_3 \wp(a(x + \omega_2)) + h_4 \wp(a(x + \omega_1 + \omega_2)), \tag{1.14}$$

where all constants $\{h\}$ are arbitrary.

In the case of interest for physics, $\tilde{V}_{12} = 0$, all solutions can be found for a "truncated" functional equation

$$V_{12}(x_1 - x_2)[W_2''(x_2) - W_1''(x_1)] + 2V_{12}''(x_1 - x_2)[W_2(x_2) - W_1(x_1)]$$
$$- 3V_{12}'(x_1 - x_2)[W_2'(x_2) + W_1'(x_1)]. \tag{1.15}$$

Indeed, one may introduce new variables into (1.12),

$$\tau = \frac{1}{2}(x_1 - x_2), \quad \rho = -\frac{1}{2}(x_1 + x_2),$$

and new unknown functions connected with W_1, W_2, V_{12} by

$$W_1(x_1) = \frac{d}{d\tau} N(\tau - \rho), \quad W_2(x_2) = \frac{d}{d\tau} L(-\tau - \rho),$$
$$V_{12}(x_1 - x_2) = \left[\frac{d\eta(\tau)}{d\tau}\right]^{-2}. \tag{1.16}$$

The functional equation (1.15) can be triply integrated, which reduces it to the functional equation without derivatives:

$$L(\tau + \rho) - N(\tau - \rho) = c_1(\rho)\eta^2(\tau) + c_2(\rho)\eta(\tau) + c_3(\rho) \tag{1.17}$$

with c_1, c_2, c_3 arbitrary functions of ρ. Expanding both sides of (1.17) in a power series in ρ, we arrive at an infinite system of ordinary differential equations

$$L(\tau) - N(\tau) = c_1(0)\eta^2(\tau) + c_2(0)\eta(\tau) + c_3(0),$$
$$(d/d\tau)^k[L(\tau) - (-1)^k N(\tau)]$$
$$= c_1^{(k)}(0)\eta^2(\tau) + c_2^{(k)}(0)\eta(\tau) + c_3^{(k)}(0), \quad k = 1, 2, \ldots$$

For $L(\tau) \neq N(\tau)$, one can obtain an equation for the function $\eta(\tau)$ by eliminating $L(\tau)$ and $N(\tau)$ from the first, third and fifth equations of the system; for $L(\tau) = N(\tau)$, one uses the second, fourth and sixth equations. Upon these simple calculations we find that solutions to (1.17) do exist provided that the function $\eta(\tau)$ satisfies one of the two equations

$$\eta'^2 = d_1(\eta - \eta_0)^2 + d_2(\eta - \eta_0) + d_3,$$

or

$$\eta'^2 = d_1(\eta - \eta_0)^2 + d_2 + d_3(\eta - \eta_0)^{-2}.$$

Thus, there exist two sets of solutions of (1.17) for which $\eta(\tau) = a \cosh(\beta\tau + \gamma) + b$ or $\eta(\tau) = [a \cosh(\beta\tau + \gamma) + b]^{1/2} + \eta_0$, where $a, b,$ β, γ, η_0 are arbitrary constants. Equation (1.15) also has only two sets of solutions:

$$W_1(x_1) = \lambda_1 \cosh(2\beta x_1 + \gamma_1) + \lambda_2 \cosh(\beta x_1 + \gamma_2),$$
$$W_2(x_2) = \lambda_1 \cosh(2\beta x_2 + \gamma_1) + \lambda_2 \cosh(\beta x_2 + \gamma_2),$$
$$V_{12}(x_1 - x_2) = \lambda_3 \left[\sinh\left(\frac{1}{2}\beta(x_1 - x_2)\right)\right]^{-2}, \qquad (1.18a)$$

and

$$W_1(x_1) = \lambda_1 \cosh(\beta x_1 + \gamma_1), \quad W_2(x_2 = \lambda_1 \cosh(\beta x_2 + \gamma_1),$$
$$V_{12}(x_1 - x_2) = \lambda_2 \left[\sinh\left(\frac{1}{2}\beta(x_1 - x_2)\right)\right]^{-2}$$
$$+ \lambda_3 \left[\sinh\left(\frac{1}{4}\beta(x_1 - x_2)\right)\right]^{-2}. \qquad (1.18b)$$

Note that the functions (1.18a) have important rational degenerations

$$W_1(x) = W_2(x) = \beta_1 x^4 + \beta_2 x^2 + \beta_3 x, \quad V_{12}(x) = \lambda x^{-2}, \quad (1.19)$$

where $\beta, \lambda, \lambda_1, \lambda_2, \lambda_3, \gamma_1, \gamma_2$ are arbitrary constants. By a limiting procedure, one may obtain from the solutions of (1.18a) both the Olshanetsky–Perelomov and the Adler system (1.3) at $n = 2$.

Particular solutions to the initial functional equation (1.12) may be pointed out, which have a structure similar to (1.18a) and (1.18b):

$$W_1(x) = W_2(x) = \lambda_1 \cosh \beta x, \quad V_{12}(x) = \frac{\lambda_2}{\sinh^2(\beta x/2)}$$

$$+ \frac{\lambda_3}{\sinh^2(\beta x/4)},$$

$$\tilde{V}_{12}(x) = \lambda_4/\sinh^2(\beta x/2) + \lambda_5/\sinh^2(\beta x/4), \tag{1.20a}$$

$$W_1(x) = W_2(x) = \lambda_1 \cosh \beta x + \lambda_2 \cosh(2\beta x),$$

$$V_{12}(x) = \lambda_3 \sinh^2(\beta x/2), \quad \tilde{V}_{12}(x) = \lambda_4 \sinh^2(\beta x/2). \tag{1.20b}$$

The existence of solutions to the general equation (1.12) different from (1.13), (1.18)–(1.20) remain still an open problem. An interesting question arises: which systems defined by potentials above admit an extension to systems of n interacting particles of equal mass interacting also with one particle of infinite mass? It seems that it might be solved only with the use of Lax pairs technique, as it will be demonstrated in Section 1.2.

1.2. The Motion of Classical Integrable Particle Systems in an External Field

In the review paper by Olshanetsky and Perelomov, the classical integrable systems of particles of equal mass (1 for convenience) were described in detail. The corresponding Hamiltonian is

$$H_0 = \sum_{i=1}^{N} \frac{1}{2} p_{i^2} + \sum_{i>j}^{N} V(x_i - x_j). \tag{1.21}$$

The Lax pairs for these systems were constructed in many ways. Our goal in this section is to use the Lax method so as to extend the part of the results of the previous section for the many-particle case.

The starting point is the equivalence of equations of motion defined by the Hamiltonian H_0 and the matrix equality

$$\frac{d\mathbf{l}}{dt} = \{\mathbf{l}, H_0\}_P = [\mathbf{l}, \mathbf{m}], \tag{1.22}$$

where l and m are $N \times N$ matrices, $\{., .\}_P$ is the Poisson bracket, $[l, m] = lm-ml$. According to (1.22), the eigenvalues of l do not depend on time t and represent a set of N independent integrals of motion. The structure of matrices l and m constituting the Lax pair was first obtained by Calogero and Moser in 1975:

$$l_{jk} = p_j \delta_{jk} + (1 - \delta_{jk}) R(x_j - x_k),$$

$$m_{jk} = m_{jj} \delta_{jk} + (1 - \delta_{jk}) R'(x_j - x_k). \tag{1.23}$$

The function $R(x)$ related to the potential V by $V(x) = R(x)R(-x)+$ const may assume values from the following set:

$$R(x) = ia/x, \, ia/\sinh\left(\frac{1}{2}\beta x\right), \, ia \coth\left(\frac{1}{2}\beta x\right),$$

$$R(x) = ia\,\frac{\sigma(x + \alpha)}{\sigma(x)}, \tag{1.24}$$

where a and β are arbitrary parameters, $i^2 = -1$, σ is the Weierstrass σ function and α is the so-called spectral parameter which does not influence $V(x)$.

If the motion of a system is determined by the Hamiltonian $H = H_0 + H_{int}$, where

$$H_{int} = \sum_{j=1}^{N} W(x_j), \tag{1.25}$$

it is then not difficult to calculate the derivative of the matrix l with respect to time:

$$\frac{dl}{dt} = \{l, H_0 + H_{int}\}_P = [l, m] - w', \tag{1.26}$$

where w' is a diagonal matrix: $w'_{jk} = \delta_{jk} W'(x_j)$.

Consider now $2N \times 2N$ matrices $\mathbf{L}$ and $\mathbf{M}$ with the block structure

$$\mathbf{L}_{jk} = l_{jk}, \, \mathbf{L}_{j+N,k} = q_{jk}, \, \mathbf{L}_{j,k+N} = q_{jk}, \, \mathbf{L}_{j+N,k+N} = -l_{jk};$$

$$\mathbf{M}_{jk} = m_{jk}, \, \mathbf{M}_{j+N,k} = -s_{jk}, \, \mathbf{M}_{j,k+N} = s_{jk}, \, \mathbf{M}_{j+N,k+N} = m_{jk}, \tag{1.27}$$

where $\mathbf{q}$ and $\mathbf{s}$ are diagonal $N \times N$ matrices:

$$q_{jk} = \delta_{jk} Q(x_j), \quad s_{jk} = \delta_{jk} S(x_j).$$

From (1.27) it follows that the Lax relation that allows us to get further integrals of motion,

$$\frac{d\mathbf{L}}{dt} = [\mathbf{L}, \mathbf{M}], \tag{1.27a}$$

holds, provided that the matrix equalities

$$\frac{d\mathbf{l}}{dt} = [\mathbf{l}, \mathbf{m}] - \{\mathbf{s}, \mathbf{q}\}, \tag{1.28a}$$

$$\frac{d\mathbf{q}}{dt} = [\mathbf{q}, \mathbf{m}] + \{\mathbf{s}, \mathbf{l}\}, \tag{1.28b}$$

are fulfilled ($\{\mathbf{a}, \mathbf{b}\}$ is the anticommutator of matrices $\mathbf{a}, \mathbf{b}$).

Comparing (1.28a) with (1.26), we find the relation between functions Q, S, W:

$$2Q(x)S(x) = W'(x).$$

Calculating the derivative of the matrix $\mathbf{q}$ and allowing for the structure of $\mathbf{q}, \mathbf{s}, \mathbf{l}, \mathbf{m}$, we obtain that the equality (1.28b) is equivalent to the following equations for Q, S, and R:

$$[Q(x) - Q(y)]R'(x - y) + R(x - y)[S(x) + S(y)] = 0, \tag{1.29a}$$

$$Q'(x) = 2S(x). \tag{1.29b}$$

Eliminating S from (1.29b), we conclude that the isospectral condition for the matrix $\mathbf{L}$ holds if $Q(x)$ obeys the functional equation

$$2[Q(x) - Q(y)]R'(x - y) + R(x - y)[Q'(x) + Q'(y)] = 0, \tag{1.30}$$

and

$$W(x) = \frac{1}{2}Q^2(x) + \text{const.} \tag{1.31}$$

It is not difficult to obtain solutions to (1.30) with $R(x)$ given by (1.24). These solutions are non-trivial only for $R(x) = a/x$,

$Q(x) = bx$ (we then arrive to the Olshanetsky–Perelomov system) and for

$$R(x) = a/x, \quad Q(x) = bx^2 + cx + d,$$

$$R(x) = a/\sinh\left(\frac{1}{2}\beta x\right), \quad Q(x) = \gamma_1 \cosh\beta x$$

$$+ \gamma_2 \sinh(\beta x + \delta) + \rho, \tag{1.32}$$

where $a, b, c, d, \gamma_1, \gamma_2, \beta, \delta, \rho$ are arbitrary constants. According to (1.31), (1.32), and (1.27a), the systems with the Hamiltonian $H_0 + H_{int}$ defined by the sets

$$V(x) = a^2/x^2, \quad W(x) = b_1 x^4 + b_2 x^3 + b_3 x^2 + b_4 x,$$

$$V(x) = a^2/\sinh^2\left(\frac{1}{2}\beta x\right),$$

$$W(x) = \gamma_1 \cosh(2\beta x) + \gamma_2 \cosh(\beta x + \delta) \tag{1.33}$$

have N independent integrals of motion:

$$I_k = (1/2k)\mathrm{Tr}((\mathbf{L})^{2k}), \quad k = 1, \ldots, N. \tag{1.34}$$

Due to the matrix equalities (1.26) and (1.28b) being valid, one may use yet another way to construct the class of N-particle system interacting with external field with another Lax pair. In fact, from (1.26), (1.28b), and (1.29), it follows that

$$\frac{1}{2}\frac{d(\mathbf{l})^2}{dt} = \frac{1}{2}[(\mathbf{l})^2, \mathbf{m}] - \frac{1}{2}\{\mathbf{l}, \mathbf{w}'\}, \tag{1.35a}$$

$$\frac{d\mathbf{q}}{dt} = [\mathbf{q}, \mathbf{m}] + 1/2\{\mathbf{l}, \mathbf{q}'\}, \tag{1.35b}$$

Summing (1.35a) and (1.35b), we get

$$d/dt\left(\frac{1}{2}\mathbf{l}^2 + \mathbf{q}\right) = \left[\frac{1}{2}\mathbf{l}^2 + \mathbf{q}, \mathbf{m}\right] + \frac{1}{2}\{\mathbf{l}, \mathbf{q}' - \mathbf{w}'\}.$$

It is easy to see that under the condition $W(x) = Q(x)$ the pair of $(N \times N)$ matrices $\tilde{\mathbf{L}} = \frac{1}{2}\mathbf{l}^2 + \mathbf{q}$ and $\mathbf{M} \cong \mathbf{m}$ satisfies Lax relation, i.e. because of a solution to (1.30) of the type (1.32), the set

$$V(x) = a^2/\sinh^2(\beta x/2), \quad W(x) = \gamma \cosh(\beta x + \delta), \tag{1.36}$$

also defines an N-particle system with the Hamiltonian $H_0 + H_{int}$ having N independent integrals of motion

$$I_k = k^{-1}\mathrm{Tr}(\tilde{L}^k), \quad k = 1, \ldots, N.$$

Note that the Adler system is a particular case of (1.36) when $\delta \to +\infty, \gamma e^{\delta} \to$ const. It is also easy to see that the potential of external field in (1.36) is no more than a particular case of the general formula (1.33).

As in the course of motion of particles in systems (1.33) and (1.36), simple asymptotic states do not appear, the study of the Poisson brackets of the integrals of motion (1.34) $\{I_k, I_l\}_P$ requires the use of algebraic methods (if these quantities vanish, the corresponding dynamical systems are completely integrable). This study will be carried out in the next section.

1.3. New Completely Integrable Multiparticle Dynamical Systems

We shall start from Equation (1.30) and the Calogero–Moser equation

$$R_{jl}R'_{jk} + R_{jk}R'_{jl} = R_{kl}(T_{jl} - T_{jk}),$$
$$R_{jk} = R(x_j - x_k), \ (T_{jl} = T(x_j - x_l)). \quad (1.37)$$

Note that $R(x)$ is an odd function, and $R'(x)$ and $T(x)$ are even functions

$$R_{jk} = -R_{kj}, \ R'_{jk} = R'_{kj}, \ T_{jk} = T_{kj}. \quad (1.38)$$

Our goal is to prove that

$$\{I_k, I_n\}_P = \sum_{j=1}^{N} \left(\frac{\partial I_k}{\partial p_j}\frac{\partial I_n}{\partial x_j} - \frac{\partial I_k}{\partial x_j}\frac{\partial I_n}{\partial p_j} \right) = 0, \quad (1.39)$$

where $\{I_k\}$ are integrals of motion defined in the previous section. We shall restrict ourselves to the even potentials H_{int},

$$W(x) = \gamma_1 \cosh(2\beta x) + \gamma_2 \cosh\beta x, \ W(x) = a_1 x^4 + a_2 x^2, \quad (1.40)$$

leaving to the reader the common case of non-even H_{int} of the general form (1.33).

Unlike the system considered by Adler, the systems with potentials (1.40) do not have simple asymptotic states: light particles are all bound by the particle of infinite mass. A more effective method is an algebraic method of studying the Poisson brackets proposed by Perelomov. We present here main steps of the proof for the potentials (1.40).

Consider eigenvalues of the Hermitian matrix $\mathbf{L}$, the algebraic functions of which are integrals $\{I_k\}$. If the Poisson brackets of any pair of these eigenvalues $\{\lambda, \omega\}$ are zero, then the integrals are in involution. Let

$$\varphi = \begin{pmatrix} \varphi_1 \\ \varphi_2 \end{pmatrix},$$

and

$$\chi = \begin{pmatrix} \chi_1 \\ \chi_2 \end{pmatrix},$$

be normalized eigenvectors of $\mathbf{L}$ corresponding to eigenvalues λ and ω ($\varphi_1, \varphi_2, \chi_1, \chi_2$ are N-dimensional vectors). By definition,

$$\mathbf{L}\varphi = \lambda\varphi,$$

$$\mathbf{L}\chi = \omega\chi,$$

or

$$\sum_{k \neq j} l_{jk}\varphi_{1k} + Q(x_j)\varphi_{2j} = \lambda\varphi_{1j},$$

$$-\sum_{k \neq j} l_{jk}\varphi_{2k} + Q(x_j)\varphi_{1j} = \lambda\varphi_{2j},$$

$$\sum_{k \neq j} l_{jk}\chi_{1k} + Q(x_j)\chi_{2j} = \omega\chi_{1j}, \tag{1.41}$$

$$-\sum_{k \neq j} l_{jk}\chi_{2k} + Q(x_j)\chi_{1j} = \omega\chi_{2j}$$

(hereafter summation runs over indices from 1 to N).

Using the structure of the matrix $\mathbf{L}$ and the relations

$$\frac{\partial \lambda}{\partial x_j} = \left(\varphi, \frac{\partial L}{\partial x_j} \varphi \right),$$

$$\frac{\partial \lambda}{\partial p_j} = \left(\varphi, \frac{\partial L}{\partial p_j} \varphi \right),$$

we reperesent the Poisson brackets $\{\lambda, \omega\}_P$ as

$$\{\lambda, \omega\}_P = \sum_j Q'(x_j)(Z_j + Z_j^*) + i \sum_{j,k \neq j} R'_{jk}(U_{jk} - U_{jk}^*), \quad (1.42)$$

where

$$Z_j = (\varphi_{1j}^* \chi_{1j}^* + \varphi_{2j}^* \chi_{2j}^*)(\varphi_{1j} \chi_{2j} - \varphi_{2j} \chi_{1j}),$$

$$U_{jk} = \varphi_{1j}^* \chi_{1j}^* A_{jk}^{(11)} + \varphi_{2j}^* \chi_{2j}^* A_{jk}^{(22)} - \varphi_{2j}^* \chi_{1j}^* A_{jk}^{(21)}$$

$$- \varphi_{1j}^* \chi_{2j}^* A_{jk}^{(12)}, \quad (1.43)$$

$$A_{jk}^{(\alpha\gamma)} = \varphi_{\alpha j} \chi_{\gamma k} - \varphi_{\alpha k} \chi_{\gamma j}, \quad \alpha, \gamma = 1, 2. \quad (1.44)$$

Since the formulas (1.41) give rise to the relations

$$\varphi_{1j} \chi_{1j} = -i(\lambda - \omega)^{-1} \sum_{\nu \neq j} R_{j\nu} A_{j\nu}^{(11)}$$

$$+ (\lambda - \omega)^{-1} Q(x_j)(\chi_{1j} \varphi_{2j} - \chi_{2j} \varphi_{1j}), \quad (1.45)$$

$$\varphi_{1j} \chi_{2j} = -i(\lambda + \omega)^{-1} \sum_{\nu \neq j} R_{j\nu} A_{j\nu}^{(12)}$$

$$+ (\lambda + \omega)^{-1} Q(x_j)(\varphi_{1j} \chi_{1j} + \varphi_{2j} \chi_{2j})$$

and two relations for $\varphi_{2j} \chi_{2j}, \varphi_{2j} \chi_{1j}$ following from Equation (1.45) by the changes of indices $(1 \to 2)$, the second term in Equation (1.42) may be transformed as follows:

$$i \sum_{j,k \neq j} R'_{jk}(U_{jk} - U_{jk}^*) = i \sum_{j,k \neq j} R'_{jk} Q(x_j)(Y_{jk} - Y_{jk}^*)$$

$$- \sum_{j,k \neq j, l \neq j} R_{kl}(T_{jl} - T_{jk}) \left[(\lambda - \omega)^{-1} \right.$$

$$\times \left(A_{jk}^{(11)} A_{jl}^{*(11)} - A_{jk}^{(22)} A_{jl}^{*(22)} \right)$$

$$+ (\lambda + \omega)^{-1} \left(A_{jl}^{(21)} A_{jl}^{*(21)} - A_{jl}^{(12)} A_{jk}^{*(12)} \right) \Big],$$

$$(1.46)$$

where

$$Y_{jk} = (\lambda - \omega)^{-1} \left(A_{jk}^{(11)} - A_{jk}^{(22)} \right) (\chi_{1j}^* \varphi_{2j}^* - \chi_{2j}^* \varphi_{1j}^*)$$

$$- (\lambda + \omega)^{-1} \left(A_{jk}^{(21)} - A_{jk}^{(12)} \right) (\chi_{1j}^* \varphi_{1j}^* + \chi_{2j}^* \varphi_{2j}^*), \quad (1.47)$$

and terms of the type $R_{jk} R'_{j\nu} + R_{j\nu} R'_{jk}$ are transformed with the use of the functional Equation (1.37). Summing in the last term of Equation (1.46) over one of the indices l or k and allowing for the symmetry properties (1.38) and the relations

$$i \sum_{k \neq l} R_{lk} A_{jk}^{(11)} = -p_l A_{jl}^{(11)} - Q(x_l)(\varphi_{1j} \chi_{2l} - \varphi_{2j} \chi_{1j})$$

$$+ \omega \varphi_{1j} \chi_{1l} - \lambda \varphi_{1l} \chi_{1j}, \quad (1.48)$$

we arrive finally at the expression for the second term in the Poisson bracket (1.42):

$$i \sum_{j,k \neq j} R'_{jk}(U_{jk} - U_{jk}^*) = i \sum_{j,k \neq j} R'_{jk} Q(x_j)(Y_{jk} - Y_{jk}^*)$$

$$+ i \sum_{l \neq j} T_{jl} Q(x_l)(X_{jl} - X_{jl}^*), \quad (1.49)$$

$$X_{jl} = (\lambda - \omega)^{-1} [A_{jl}^{*(11)} (\varphi_{1j} \chi_{2l} - \varphi_{2l} \chi_{1j})$$

$$+ A_{jl}^{*(22)} (\varphi_{2j} \chi_{1l} - \varphi_{1l} \chi_{2j})] + (\lambda + \omega)^{-1}$$

$$\times [A_{jl}^{*(12)} (\varphi_{1j} \chi_{1l} + \varphi_{2l} \chi_{2j})$$

$$+ A_{jl}^{*(21)} (\varphi_{2j} \chi_{2l} + \varphi_{1l} \chi_{1j})].$$

Note that all the terms in the right-hand side of Equation (1.46) having zero order in $Q(x_j)$ and $Q(x_\nu)$, vanish because of the symmetry properties of R_{jk}, R'_{jk}, $A_{jk}^{(\alpha\gamma)}$.

Using the equalities resulting from Equation (1.45)

$$\varphi_{1j}\chi_{1j} + \varphi_{2j}\chi_{2j} = -i(\lambda - \omega)^{-1} \sum_{l \neq j} R_{jl}\left(A_{jl}^{(11)} - A_{jl}^{(22)}\right),$$

$$\varphi_{1j}\chi_{2j} - \varphi_{2j}\chi_{1j} = -i(\lambda + \omega)^{-1} \sum_{l \neq j} R_{jl}\left(A_{jl}^{(12)} + A_{jl}^{(21)}\right),$$

we transform the first term of Equation (1.42) and the expression for Equation (1.47), and as a result, the Poisson bracket becomes

$$\{\lambda, \omega\}_P = (\lambda^2 - \omega^2)^{-1} \sum_{j,k \neq j \neq l} [(R'_{jl}R_{jk} + R'_{jk}R_{jl})Q(x_j)$$

$$+ R_{jk}R_{jl}Q'(xj)][(A_{jl}^{(11)} - A_{jl}^{(22)})(A_{jk}^{*(12)} + A_{jk}^{*(21)}),$$

$$+ (A_{jl}^{*(11)} - (A_{jl}^{*(22)})(A_{jk}^{(12)} - (A_{jk}^{(21)})]$$

$$+ i \sum_{j,l \neq j} T_{jl}Q(x_l)(X_{jl} - X_{jl}^*). \tag{1.50}$$

The expression in braces symmetric in indices (k, l) may be transformed when $k \neq l$, using three times the functional equation (1.37) so that the result does not contain terms with $Q'(x_j), Q'(x_k), Q'(x_l)$:

$$(R'_{jl}R_{jk} + R'_{jk}R_{jl})Q(x_j) + R_{jk}R_{jl}Q'(x_j)$$

$$= Q(x_l)R_{jk}R_{lk}^{-1}(R'_{jl}R_{lk} + R_{jl}R'_{lk} + (k \leftrightarrow l). \tag{1.51}$$

Note also that for $k \neq l$, the equality holds

$$R_{jk}(R'_{jl}R_{lk} + R_{jl}R'_{lk}) = -R_{lk}^2 T_{jl}, \tag{1.52}$$

which follows from the functional equation (1.37).

Thus, with Equations (1.51) and (1.52) the Poisson bracket (1.50) may be represented in the form

$$\{\lambda, \omega\}_P = (\lambda^2 - \omega^2)^{-1} \sum_{j,k \neq j} [2R'_{jk}Q(x_j) + Q'(x_j)R_{jk}]$$

$$\times R_{jk}[(A_{jk}^{(11)} - A_{jk}^{(22)})(A_{jk}^{*(12)} + A_{jk}^{*(21)})$$

$$+ (A_{jk}^{*(11)} - A_{jk}^{*(22)})(A_{jk}^{(12)} + A_{jk}^{(21)})] + (\lambda^2 - \omega^2)^{-1}$$

$$\times \sum_{j,k \neq j, l \neq j, k \neq l} T_{jl} Q(x_l) R_{kl} [(A_{jl}^{(11)} - A_{jl}^{(22)})(A_{jk}^{*(12)} + A_{jk}^{*(21)})$$

$$+ (A_{jl}^{*(11)} - A_{jl}^{*(22)})(A_{jk}^{(12)} + A_{jk}^{(21)}) + (k \leftrightarrow l)]$$

$$+ i \sum_{j,l \neq j} T_{jl} Q(x_j)(X_{jl} - X_{jl}^*). \qquad (1.53a)$$

As the term in the bracket of the first sum of equation (1.53a) is symmetric in indices (j, k), then interchanging the latter and using Equation (1.37) one may show that this sum vanishes. Summing in the second in Equation (1.26a) over index k with the use of relation like Equation (1.48) and applying the definition of X_{jl} (Equation (1.49)), we finally get

$$\{\lambda, \omega\}_P = i \sum_{j,l \neq j} T_{jl} Q(xl)(\lambda^2 - \omega^2)^{-1}(Z_{jl} - Z_{jl}^*), \qquad (1.53b)$$

where

$$Z_{jl} = \omega \left(A_{jl}^{*(11)} A_{jl}^{(21)} + A_{jl}^{*(22)} A_{jl}^{(12)} + A_{jl}^{(11)} A_{jl}^{*(21)} + A_{jl}^{(22)} A_{jl}^{*(12)} \right)$$

$$+ \lambda \left(A_{jl}^{*(11)} A_{jl}^{(12)} + A_{jl}^{*(22)} A_{jl}^{(21)} + A_{jl}^{(11)} A_{jl}^{*(12)} + A_{jl}^{(22)} A_{jl}^{*(21)} \right).$$

However, it is easy to see that Z_{jl} are real, consequently the right-hand side of equation (1.53b) vanishes for any pairs $\{\lambda, \omega\}$ that prove the involution of the set of integrals of $\{I_k\}$.

Therefore, according to the Liouville theorem, the new systems with potentials (1.40) are completely integrable and the solutions of classical equation of motion can be in principle obtained in quadratures. Unfortunately, the Liouville theorem does not tell us how to do it, and we get these solutions partially empirically, and partially (for rational potentials (1.40)) by using highly non-trivial methods of algebraic geometry.

1.4. Integrable Particle Systems in an External Field Generated by a Particle of Infinite Mass and Motion of the Poles of the Solutions to Inhomogeneous Burgers–Hopf Equation

The methods of integration of equations of motion for the systems of light particles with equal mass were known since the work of Olshanetsky and Perelomov. When the interaction of these systems with a particle of infinite mass is switched on, the problem becomes extremely complicated and all the known methods do not work. We shall consider here some particular solutions of the systems with the Hamiltonians

$$H = \sum_{j=1}^{N}[1/2p_j^2 + W(x_j)] + \sum_{i>j}^{N} V(x_i - x_j),$$

where

$$V(x) = a/x^2, \quad W(x) = \gamma_1 x^4 + \gamma_2 x^2 + \gamma_3 x, \qquad (1.54a)$$

$$V(x) = a/\sinh^2(x/2), \quad W(x) = \gamma_1 \cosh(2x)$$

$$+ \gamma_2 \cosh x + \gamma_3 \sinh x \qquad (1.54b)$$

with some restrictions to constants $a, \gamma_1, \gamma_2, \gamma_3$ and initial conditions. Consider first the system of first-order equations

$$\frac{dx_i}{dt} = A\sum_{j\neq i}^{N} Z(x_i - x_j) + Y(x_i), \quad i = 1, \ldots, N. \qquad (1.55)$$

It can be shown that solutions of equation (1.55) satisfy the equations of motion defined by the Hamiltonian in the cases (1.54a) and (1.54b) with a particular choice of the functions $Z(\xi), Y(\xi)$, and constants in (1.54a) and (1.54b):

$$Z(\xi) = \xi^{-1}, \quad Y(\xi) = \alpha\xi^2 + \beta,$$

$$\gamma_1 = -\frac{1}{2}\alpha^2, \quad \gamma_2 = -\alpha\beta, \quad \gamma_3 = -\alpha A(N-1),$$

$$a = -A^2, \qquad (1.56a)$$

$$Z(\xi) = \coth(\xi/2), \quad Y(\xi) = \alpha \cosh \xi + \beta,$$

$$\gamma_1 = -\frac{1}{4}\alpha^2, \ \gamma_2 = -\alpha\beta, \ \gamma_3 = -\alpha A(N-1), \quad a = -A^2. \tag{1.56b}$$

Thus, the analysis of equation (1.55) allows us to find solutions to the equations of motion for the systems (1.54a) and (1.54b) with the initial conditions

$$p_i(0) = A \sum_{j \neq i}^{N} Z(x_i(0) - x_j(0)) + Y(x_i(0)). \tag{1.57}$$

The quantities $x_i(0)$ represent initial conditions for the system (1.55) and may be taken arbitrarily.

Let us now show that the system (1.55) with conditions (1.57) determines singular solutions to inhomogeneous Burgers–Hopf equations. To this end, following Calogero, we put

$$u(x,t) = \sum_{i=1}^{N} [x - x_i(t)]^{-1}, \tag{1.58a}$$

$$u(x,t) = \sum_{i=1}^{N} \coth\frac{1}{2}[x - x_i(t)]. \tag{1.58b}$$

With the use of equation (1.55) and the functional relation for functions $Z(\xi)$ of type (1.56)

$$Z(\xi)Z'(\eta) - Z'(\xi)Z(\eta) = Z(\xi + \eta)[Z'(\eta) - Z'(\xi)],$$

we may express the time derivatives of $u(x,t)$ in terms of u, u_x, u_{xx} and functions which depend only on x:

$$u_t = -\frac{1}{2}A(u_{xx} + 2uu_x) + [u\psi_1(x) + \tilde{\psi}_1(x)]_x, \tag{1.59a}$$

$$u_t = -A(u_{xx} + uu_x) + [u\psi_2(x) + \tilde{\psi}_2(x)]_x, \tag{1.59b}$$

where

$$\psi_1(x) = -\alpha x^2 - \beta, \quad \tilde{\psi}_1(x) = \alpha N x, \tag{1.60a}$$

$$\psi_2(x) = -\alpha \cosh x - \beta, \quad \tilde{\psi}_2(x) = \alpha N \sinh x. \tag{1.60b}$$

Substituting $v(x,t) = u - A^{-1}\psi_{1,2}(x)$ we transform (1.59a) and (1.59b) to the form

$$v_t + c_1 v_{xx} + c_2 v v_x = \psi(x).$$

The inhomogeneous Burgers–Hopf equation has the Lax representation and can always be reduced to a linear equation. Particle coordinates $x_i(t)$, according to (1.58), determine the position of poles of its solutions. For our purposes, it is more convenient to consider directly the system of equations (1.59a) and (1.59b). Making use of the substitution

$$u = \text{const} \times (\varphi_x/\varphi),$$

we obtain from (1.59a) and (1.59b) the linear equations for $\varphi(x,t)$:

$$\varphi_t + \frac{1}{2}A\varphi_{xx} - \varphi_x\psi_1(x) - \varphi\tilde{\psi}_1(x) = 0, \qquad (1.61a)$$

$$\varphi_t + A\varphi_{xx} - \varphi_x\psi_2(x) - \varphi\tilde{\psi}_2(x) = 0. \qquad (1.61b)$$

From (1.58a) and (1.58b) it follows that solutions to equations (1.61a) and (1.61b) should obey the initial conditions

$$\varphi(x,0) = \prod_{i=1}^{N}[x - x_i(0)], \qquad (1.62a)$$

$$\varphi(x,0) = \prod_{i=1}^{N} \sinh\frac{1}{2}[x - x_i(0)]. \qquad (1.62b)$$

The quantities $x_j(t)$ we are interested in are zeros of the functions $\varphi(x,t)$.

The solution to the Cauchy problem for equations (1.61a) and (1.61b) with initial conditions (1.62a) and (1.62b) can be found as follows. We shall look for functions $\varphi(x,t)$ in the form

$$\varphi^{(1)}(x,t) = \sum_{k=0}^{N} x^k c_k^{(1)}(t), \qquad (1.63a)$$

$$\varphi^{(2)}(x,t) = \sum_{k=0}^{N} \left(\sinh\frac{1}{2}x\right)^k \left(\cosh\frac{1}{2}x\right)^{N-k} c_k^{(2)}(t). \qquad (1.63b)$$

It can be easily verified that the corresponding functions are of the form (1.58a) and (1.58b). Substituting (1.63) into (1.61) and using the explicit expressions (1.60) for $\psi_{1,2}(x)$ and $\tilde{\psi}_{1,2}(x)$ we arrive at the systems of ordinary differential equations of the first order for $c_k^{(1,2)}(t)$. Using the notation

$$c^{(\epsilon)}(t) = \{c_0^{(\epsilon)}(t), \ldots, c^{(\epsilon)}(t)\}, \quad \epsilon = 1, 2,$$

we rewrite these systems in the form

$$dc^{(\epsilon)}/dt = T^{(\epsilon)} c^{(\epsilon)}, \tag{1.64}$$

where $T(\epsilon)$ are $(N+1) \times (N+1)$ matrices independent of time and initial values of the coordinates $\{x_i(0)\}$. Non-zero are only the elements

$$T_{00}^{(1)} = -\beta, \ T_{02}^{(1)} = -a, \ T_{N-1,N}^{(1)} = N\beta,$$

$$T_{N-1,N-2}^{(1)} = 2a, \ T_{N,N-1}^{(1)} = \alpha,$$

$$T_{k,k-1}^{(1)} = (N+1-k)\alpha, \ T_{k,k+1}^{(1)} = -\beta(k+1),$$

$$T_{k,k+2}^{(1)} = \frac{1}{2}(k+1)(k+2)a, \ 1 \le k \le N-2, \tag{1.65a}$$

$$T_{00}^{(2)} = -\frac{1}{4}Na, \ T_{01}^{(2)} = -\frac{1}{2}(\alpha+\beta), \ T_{02}^{(2)} = -\frac{1}{2}a,$$

$$T_{10}^{(2)} = \frac{1}{2}N(\alpha-\beta), \ T_{11}^{(2)} = \frac{1}{4}(1-3N)a,$$

$$T_{12}^{(2)} = -(\alpha+\beta), \ T_{13}^{(2)} = -\frac{3}{4}a,$$

$$T_{N-1,N-2}^{(2)} = (\alpha-\beta), \ T_{N-1,N-1}^{(2)} = \frac{1}{4}(1-N)a,$$

$$T_{N-1,N}^{(2)} = -\frac{1}{2}N(\alpha+\beta), \ T_{N,N-1}^{(2)} = \frac{1}{2}(\alpha+\beta),$$

$$T_{NN}^{(2)} = -\frac{1}{4}N(N+1)a,$$

$$T_{k,k-2}^{(2)} = -\frac{1}{4}(N+1-k)(N+2-k)a,$$

$$T_{k,k-1}^{(2)} = \frac{1}{2}(\alpha - \beta)(N + 1 - k),$$

$$T_{kk}^{(2)} = \frac{1}{4}a[N(1 + 2k) - k^2],$$

$$T_{k,k+1}^{(2)} = -\frac{1}{2}(k + 1)(\alpha + \beta), \quad 2 \leq k \leq N - 2. \tag{1.65b}$$

From (1.62a) and (1.62b) it also follows that $c^{(\epsilon)}(0)$ can be expressed in terms of $\{x_i(0)\}$:

$$c_{N-k}^{(1)}(0) = (-1)^k \sum_{1 \leq \lambda_1 < \ldots < \lambda_k \leq N} x_{\lambda_1}(0) \cdots x_{\lambda_k}(0),$$

$$c_N^{(1)}(0) = 1,$$

$$c_k^{(2)}(0) = (-1)^{N-k} \sum_{1 \leq \lambda_1 < \ldots < \lambda_k \leq N} \cosh\frac{1}{2}x_{\lambda_1}(0)$$

$$\times \cosh\frac{1}{2}x_{\lambda_k}(0) \prod_{\mu \neq \lambda_1,\ldots,\lambda_k}^{N} \sinh\frac{1}{2}x_\mu(0),$$

$$c_0^{(2)}(0) = (-1)^N \prod_{i=1}^{N} \sinh\frac{1}{2}x_i(0).$$

As the matrices (1.65a) and (1.65b) are constant, it is easy to find the solutions to equations (1.64):

$$c^{(\epsilon)}(t) = \exp(tT^{(\epsilon)})c^{(\epsilon)}(0).$$

Here, we also present the expressions for the functions $\varphi(x,t)$,

$$\varphi^{(1)}(x,t) = \sum_{k=0}^{N} x^k [\exp(tT^{(1)})c^{(1)}(0)]_k, \tag{1.66a}$$

$$\varphi^{(2)}(x,t) = \sum_{k=0}^{N} \left(\sinh\frac{1}{2}x\right)^k \left(\cosh\frac{1}{2}x\right)^{N-k}$$

$$\times [\exp(tT^{(2)})c^{(2)}(0)]_k. \tag{1.66b}$$

Formulae (1.65) and (1.66) give us the solution to the problem of integration of the equations of motion for systems (1.54a) and (1.54b) with initial conditions (1.57): the coordinates of the particles $x_j(t)$ are the zeros of the functions (1.66a) and (1.66b). In the same way, one may derive explicit solutions to the equations of motion for integrable classical systems which are ensembles of $2N$ light particles posed symmetrically with respect to the origin of coordinates and having opposite momenta. The Hamiltonian of these systems is similar in structure to the original one:

$$H = \sum_{i=1}^{N} \left[\frac{1}{2}p_j^2 + W(x_j) \right] + \sum_{i>j}^{N} [V(x_i - x_j) + V(x_i + x_j)],$$

where $\{V, W\}$ may assume the following forms:

$$V(x) = a/x^2, \ W(x) = \gamma_1/x^2 + \gamma_2 x^2 + \gamma_3 x^4 + \gamma_4 x^6, \quad (1.67a)$$

$$V(x) = a/\sinh^2(x/2), \ W(x) = \gamma_1/\sinh^2(x/2)$$
$$+ \gamma_2/\sinh^2 x + \gamma_3 \cosh x + \gamma_4 \cosh(2x). \quad (1.67b)$$

In particular, in the case (1.67b) and initial conditions (1.57) for

$$Y(x) = \alpha \cosh x + \beta \coth x + \gamma \coth(x/2), \quad Z(x) = \coth(x/2),$$

the coordinates of particles obey the system of first-order equations

$$\frac{dx_i}{dt} = A \sum_{j \neq i} [Z(x_i - x_j) + Z(x_i + x_j)] + Y(x_i),$$

which is a generalization of (1.55).

The constants $a, \gamma_1, \gamma_2, \gamma_3$ and γ_4 are connected with A, α, β and γ:

$$a = -A^2, \ \gamma_1 = -\frac{1}{2}\gamma(\gamma + \beta), \ \gamma_2 = -\frac{1}{2}\beta^2,$$

$$\gamma_3 = \alpha[\beta + \gamma + 2A(N-1)], \ \gamma_4 = -\frac{1}{4}\alpha^2.$$

To the above system of differential equations of first order also corresponds a singular solution of the inhomogeneous Burgers–Hopf equation, the construction of which allows one to find the functions $x_i(t)$.

The method we have used does not allow us to obtain solutions of the equations of motion from the systems (1.54a) and (1.54b) and (1.67a) and (1.67b) without the restrictions (1.57) on the initial conditions $\{p_i(0), x_i(0)\}$ and without (1.56) on the constants $(a, \gamma_1, \ldots, \gamma_4)$. In the next two sections, we shall consider another method which allows one to obtain the solutions of equations of motion without restrictions on initial conditions.

1.5. Classical Sutherland Systems in the Morse Potential

We shall start with the Hamiltonian (1.1) with slightly changed constants in the potentials $\{V, W\}$ for convenience. Namely, we shall consider (1.1) in the hyperbolic case when

$$V(x) = \frac{g^2}{\sinh^2(x)}, \quad W(x) = 8\tau^2 g^2(e^{4x} - e^{2x}). \quad (1.68)$$

The form of $W(x)$ (1.68) can be easily obtained from the general case (1.33) by shifting of the coordinate to infinity and "renormalization" of the couplings γ_1, γ_2. It means that we do not know till now how to integrate the equations of motion for the Hamiltonian (1.1) with the potential $W(x)$ of the form (1.33), and shall investigate only the limit case (1.68), i.e. the Sutherland systems interacting with a particle of infinite mass by the Morse potential. In quantum case, there might be states of discrete spectrum and continuous spectrum. In classical case, some part of light particles (or all of them) can be bound and the motion might be finite. But in most general cases, light particles can scatter from each other and from the Morse potential, moving to infinity. We shall propose a method of integration of the equations of motion for *arbitrary* initial conditions with no restriction to the parameters g and τ. It is based on the Lax representation (1.27a). We shall find the existence conditions of equilibrium N-particle configurations (when all of them are bound by the Morse potential), various characteristics of equilibrium state and their unexpectedly simple connection with the discrete spectrum of quantum systems and positions of zeros of the generalized Laguerre polynomials. We shall show that generally finite motion is not periodic.

First, let us integrate the classical equations of motion. Let us write the matrices $\mathbf{L}$ and $\mathbf{M}$ in the Lax representation (1.27a) in the form

$$\mathbf{L} = \begin{pmatrix} l & \psi \\ \psi & -l \end{pmatrix},$$

$$\mathbf{M} = \begin{pmatrix} m & \psi'/2 \\ -\psi'/2 & m \end{pmatrix},$$

where l, m and ψ, ψ' are $N \times N$ matrices with the structure

$$l_{jk} = p_j\delta_{jk} + (1 - \delta_{jk})ig(\sinh(x_j - x_k))^{-1},$$

$$m_{jk} = -ig(1 - \delta_{jk})\coth(x_j - x_k)(\sinh(x_j - x_k))^{-2}$$
$$+ i\delta_{jk}g\sum_{s\neq j}(\sinh(x_s - x_j))^{-2}$$

$$\psi_{jk} = \delta_{jk}4\tau g\left(e^{2x_j} - \frac{1}{2}\right), \tag{1.69}$$

$$\psi'_{jk} = \frac{\partial\psi_{jk}}{\partial x_j} = 2\psi_{jk} + 4\tau g\delta_{jk}, \quad 1 \le j, k \le N.$$

It may be easily verified that the matrix equation (1.27a) is equivalent to the Hamilton equation of motion,

$$\frac{dp_j}{dt} = -\frac{\partial H}{\partial x_j}, \quad \frac{dx_j}{dt} = p_j,$$

which owing to (1.27a) possesses N constants of motion $I_k = \mathrm{Tr}(\mathbf{L}^{2k}), k = 1, \ldots, N$.

Explicit integration of the equations of motion require, however, additional N variables of the "angle" type canonically conjugate to $\{I_k\}$. Adler has found a more simple method of solution that allows one to reduce the equations of motion to the Riccatti equation. For the potential (1.68) the Adler method does not work, and more effective is a slightly modified other method used earlier by Olshanetsky and Perelomov for integration of the Sutherland systems with $W = 0$.

Consider a $(2N \times 2N)$ matrix of rank N

$$X = \begin{pmatrix} 0 & 0 \\ 0 & z \end{pmatrix}, \quad z_{jk} = z(x_j)\delta_{jk}.$$

Computing the matrix commutator $[X, \mathbf{M}]$ and anticommutator $\{X, \mathbf{L}\}$, we get

$$Y = -\frac{dX}{dt} + [X, \mathbf{M}] + \{X, \mathbf{L}\}$$

$$= \begin{pmatrix} 0 & z\left(\psi - \frac{\psi'}{2}\right) \\ z\left(\psi - \frac{\psi'}{2}\right) & -\frac{dz}{dt} + [z, m] - \{z, l\} \end{pmatrix}. \tag{1.70}$$

It is easy to see that the matrix equation

$$\frac{dy}{dt} = [y, m] + \left\{l, \frac{y'}{2}\right\}, \tag{1.71}$$

is valid only for matrices l, m of the form (1.69) and a diagonal matrix y given by

$$y_{jk} = y(x_j)\delta_{jk},$$

$$y(x_j) = C_1 \exp(2x_j) + C_2 \exp(-2x_j) + C_3,$$

where C_1, C_2 and C_3 are arbitrary constants. As $C_1 = C_3 = 0$, we get the equality $y'/2 = -y$; inserting it into (1.71) and comparing with (1.70) we find that the right lower block of the matrix vanishes provided that

$$z(x_j) = \exp(-2x_j).$$

Further, according to (1.69), the matrix $\psi - \psi'/2 = -2\tau g E$, where E is a unit $N \times N$ matrix. Consequently, for the functions $z(x_j)$ the equality

$$Y = \begin{pmatrix} 0 & -2g\tau z \\ -2g\tau z & 0 \end{pmatrix} = -2g\tau[C, X],$$

$$C = \begin{pmatrix} 0 & E \\ -E & 0 \end{pmatrix}$$

holds, and the time dependence of the matrix X is given by the equation

$$\frac{dX}{dt} = [X, M] + \{X, \mathbf{L}\} - 2g\tau[C, X]. \tag{1.72}$$

Let us introduce in Equations (1.27a) and (1.72) the unitarity-equivalent matrices

$$\tilde{\mathbf{L}} = \Omega^{-1}\mathbf{L}\Omega, \quad \bar{X} = \Omega^{-1}X\Omega,$$

$$\Omega(0) = \begin{pmatrix} E & 0 \\ 0 & E \end{pmatrix}, \quad \frac{d\Omega}{dt}\Omega^{-1} = \mathbf{M}.$$

Since $[C, \tilde{\mathbf{M}}] = 0$, C is an invariant of the transformation given by the matrix $\Omega : \tilde{C} = C$. Equations (1.27a) and (1.72) no longer contain $\mathbf{M}$:

$$\frac{d\tilde{\mathbf{L}}}{dt} = 0, \quad \frac{d\tilde{X}}{dt} = \{\tilde{X}, \tilde{\mathbf{L}}\} - 2g\tau[C, \tilde{X}]. \tag{1.73}$$

Solution of these equations is obvious: from the first equation we get that $\tilde{\mathbf{L}} = \text{const} = \mathbf{L}(0)$; with that condition the second is a linear equation with constant coefficients and the general solution is

$$\tilde{X}(t) = \exp(\Lambda t)\tilde{X}(0)\exp(\Lambda^{\dagger}t), \tag{1.74}$$

where, owing to (1.69) and (1.72),

$$\Lambda = \begin{pmatrix} l(0) & \chi(0) - 4\tau g E \\ \chi(0) & -l(0) \end{pmatrix}, \quad \chi_{jk}(0) = 4\tau g \delta_{jk} e^{2x_j(0)}, \tag{1.75}$$

and $\Lambda^{\dagger}$ is a matrix Hermitian conjugate to Λ. Thus, the problem of integration of the equations of motion of the Sutherland system in the Morse potential is solved: coordinates of particles are given by the expressions

$$x_j(t) = -\frac{1}{2}\log(\tilde{z}_j(t)),$$

where $\tilde{z}_j(t)$ are either non-zero eigenvalues of $\tilde{X}(t)$ (Equation (1.74)) or eigenvalues of the matrix $v(t) = x(0)\tilde{v}(t)$, where $\tilde{v}(t)$ is the right lower block of the $(2N \times 2N)$-matrix $\exp(\Lambda^{\dagger}t \exp(\Lambda t)X(0)$. According to Equation (1.75), Λ is completely determined with the initial velocities of particles and their positions.

Now, let us turn to the qualitative analysis of trajectories. There exist $(N+1)$ asymptotic states different in structure in which n particles, as time passes, go to infinity, whereas the remaining $(N-n)$ ones perform finite motion in the potential created by a particle of infinite mass, $0 \leq n \leq N$.

Which of the states will be occupied by the N-particle system may in principle be solved by studying the structure of the Λ-matrix. When all the eigenstates of Λ are real, quasiperiodic motion is impossible, and after a number of particles have gone to infinity, the others should in time occupy their equilibrium positions. And vice versa, when all the roots of the equation $\det(\lambda I - \Lambda) = 0$ are imaginary (I is the unit $2N \times 2N$ matrix), the motion can be only finite (note that the squared eigenvalues of Λ are always real since the matrix Λ^2 is Hermitian). We cannot, however, point to the criterion which would allow a direct determination of the asymptotic state from the initial conditions. Only for a two-particle system can it be shown that there are no inelastic processes with knock-out or capture into a bound state. Processes involving more than two particles require a somewhat more labor-consuming study.

Now, let us study the equilibrium point of our systems more carefully. As the Morse potential decreases at infinity very rapidly, it is clear that a system of an infinite number of particles repelled by the law (1.68) cannot be in the equilibrium state in it. In other words, the constant τ defining the depth of the potential well cannot be as small as desired for equilibrium configurations of a given nimber of light particles, N. In quantum mechanics this situation is characterized by the existence condition of the discrete spectrum ($\hbar = 1$)

$$\tau g - \left(\frac{1}{2} + \sqrt{g^2 + \frac{1}{4}(N-1)} \right) > \frac{1}{2},$$

as it will be shown in the next chapter.

There exists an analog of this inequality for equilibrium states of classical systems. It is known that for particles repelled by the law $1/x^2$ or $-\log|x|$ in the potential of a linear oscillator, the coordinates in equilibrium coincide with the zeros of Hermite polynomials from the pioneering work of Calogero. Let us show that for the Sutherland systems in the Morse potential the coordinates of particles in equilibrium are related to the zeros of generalized Laguerre polynomials, which will allow us to find a classical analog of a quantum restriction.

The equilibrium equations $\frac{\partial H}{\partial x_j} = 0$ are of the form

$$-\sum_{k \neq j} \frac{\cosh(x_j - x_k)}{\sinh^2(x_j - x_k)} + 8\tau^2 e^{2xj}(2e^{2xj} - 1) = 0. \qquad (1.75a)$$

The left-hand side of this equation is a rational function of the parameters $y_j = e^{2xj}$:

$$-\sum_{k \neq j} \frac{y_k(y_i + y_k)}{(y_j - y_k)^2} + 2\tau^2(2y_j - 1) = 0. \qquad (1.76)$$

The most simple way to calculate the sum in (1.76) is the way proposed by Ahmed. Consider an integral over a contour in the complex y-plane, enveloping all the points with coordinates obeying the system (1.76), of the function

$$F_j(y) = \frac{y(y_j + y)p_N'(y)}{(y - y_j)^3 p_N(y)}.$$

Here, $p_N(y)$ is a polynomial of degree N of the type $\prod_{m=1}^{N}(y - y_m)$ and the set of numbers $\{y_m\}$ are solutions to Equation (1.76). As the function $F_j(y)$ decreases as y^{-2} when $|y| \to \infty$, the sum of residues of $F_j(y)$ at all its poles should vanish. At the points $\{y_k\}$ with $k \neq j$, the poles of $F_j(y)$ are simple, and the residues at those poles are

$$\mathrm{res}F_j(y)|_{y=y_k} = \frac{y_k(y_k + y_j)}{(y_k - y_j)^3}. \qquad (1.77)$$

Now, let us calculate the residue of $F_j(y)$ at the point $y = y_j$ with a fourth-order pole. Setting $y = y_j + \zeta$, we get

$$\mathrm{res}F_j(y)|_{y=y_j} = \frac{1}{6}\frac{d^3}{d\zeta^3}\left[\zeta(y_j + \zeta)(2y_j + \zeta)\frac{p_N'(\zeta + y_j)}{p_N(\zeta + y_j)}\right]|_{\zeta=0}.$$

Then, expanding $p_N(y_j + \zeta)$ up to fourth-order terms in ζ and remembering the equality $p_N(y_j) = 0$, we obtain

$$\mathrm{res}F_j(y)|_{y=y_j} = \frac{1}{4}(2a + y_j(4b - 3a^2) + y_j^2(a^3 - 2ab + c)), \qquad (1.78)$$

where the notation is

$$a = \frac{p_N''(y_j)}{p_N'(y_j)}, \quad b = \frac{p_N'''(y_j)}{p_N'(y_j)}, \quad c = \frac{p_N^{IV}(y_j)}{p_N'(y_j)}. \tag{1.78a}$$

We assume that $p_N(y)$ obeys a second-order differential equation of the type

$$A(y)p_N''(y) + B(y)p_N'(y) + \lambda_N p_N(y) = 0. \tag{1.79}$$

Subsequent differentiation of (1.79) expresses the entities a, b and c in terms of $A(y_j), B(y_j)$, their derivatives, and the constant λ_N, as follows:

$$a = \frac{B}{A}, \quad b = -A^2[B(B + A') - A(\lambda_N + B')],$$

$$c = A^{-1}[A(\lambda_N + B')(B + 2A') + 2(B' + A'' + \lambda_N)AB$$

$$-A^2 B'' - B(B + A')(B + 2A')].$$

From comparison of these equations with (1.78) it follows that the residue of $F_j(y)$ at the point $y = y_j$ can be a linear function of y_j only when $A = y_j$. Upon some transformations we get for the residue the simple expression

$$\text{res} F_j(y)|_{y=y_j} = \frac{1}{4}[B'(y_j)B(y_j) - 2B'(y_j) - 2\lambda_N - y_j B''(y_j)].$$

The equality to zero of the sum of all the residues of $F_j(y)$ is equivalent to Equation (1.76) under the condition

$$B'(y_j)B(y_j) - 2B'(y_j) - 2\lambda_N - y_j B''(y_j) = 8\tau^2(2y_j - 1).$$

From this equation it is not difficult to determine $B(y_j)$:

$$B(y_j) = -4\tau y_j + 2\left(\tau + 1 - \frac{\lambda_N}{4\tau}\right).$$

Equation (1.79) has a solution of the polynomial type of degree N if the constant λ_N is of the form

$$\lambda_N = 4N\tau.$$

This solution is obviously the generalized Laguerre polynomial

$$p_N(y) = L_N^{(\gamma)}(4\tau y), \quad \gamma = 2(\tau - N) + 1. \tag{1.80}$$

Thus, we have arrived at the following conclusion: the coordinates of particles of our systems in equilibrium are determined by the roots $\{\alpha_j\}$ of the above generalized Laguerre polynomial:

$$x_j = \frac{1}{2} \log\left(\frac{\alpha_j}{4\tau}\right). \tag{1.80a}$$

According to these relations, for equilibrium configuration to exist it is necessary that all the roots of $L_N^{(\gamma)}$ be real and positive. From the theory of classical orthogonal polynomials it is known that this condition is fulfilled with $\gamma > -1$; only in this case for any integer $m > 0$ are the functions $L_m^{(\gamma)}(y)$ orthogonal with a certain weight in the interval $(0, \infty)$, and all the roots of $L_m^{(\gamma)}(y)$ are inside that interval. Thus, the points of equilibrium of our dynamic systems exist only under the condition

$$\tau > N - 1.$$

Note also that simple algebraic expressions known for the coefficients of the Laguerre polynomials make possible the calculation of any polynomials symmetric with respect to permutations of $\{y_j\}$. Specifically, we can cite the explicit expression for the coordinate of the center of mass of a particle system in equilibrium:

$$Q = \frac{1}{N} \sum_{j=1}^{N} x_j = \frac{1}{2N} \log\left[(4\tau)^{-N} \prod_{j=1}^{N} y_j \right]$$

$$= \frac{1}{2N} \left[\log \frac{\Gamma(2\tau - N + 2)}{\Gamma(2\tau - 2N + 2)} - N \log 4\tau \right],$$

where $\Gamma(\zeta)$ is the Euler Γ-function. When $\tau \to N - 1$, the expression for Q, in accordance with the above formula for restriction to τ, is divergent, which corresponds to going of one of the particles to infinity under the action of repulsion forces of other particles.

The connection (1.80) and (1.80a) between the solutions to the system of Equations (1.76) and the roots if polynomials (1.80) can be used to determine the dependence of the energy of the equilibrium state and frequencies of small oscillations nearby on the parameter τ.

The energy is given by

$$E = 4g^2(S_0 + 2\tau^2(S_2 - S_1)),$$

$$S_0 = \sum_{j=1}^{N} \sum_{k \neq j}^{N} \frac{y_j y_k}{(y_j - y_k)^2}, \quad S_\alpha = \sum_{j=1}^{N} y_j^\alpha, \quad \alpha = 1, 2,$$

where y_j obey the Equation (1.76). Now we shall express the double sum S_0 through S_1 and S_2. To this end, consider an integral over a fairly distant closed contour of the function

$$F(y) = \sum_{j=1}^{N} \frac{y y_j}{(y - y_j)^2} \frac{p'_N(y)}{p_N(y)}$$

and separately compute the residues at the points $y_k(k \neq j)$ and y_j. The first sum equals $2S_0$; the other may be transformed by using the equalities (1.78a), which gives

$$S_0 = -\frac{2}{3}\tau^2 S_2 - \frac{\tau}{3} S_1(1 - 2\tau) - \frac{N(\tau - N)(\tau - N + 1)}{6}.$$

Multiplying (1.76) by y_j and summing over the index we obtain, owing to antisymmetry of terms in the resulting double sum, the simple relation

$$S_2 - 2S_1 = 0.$$

The quantity S_1 is the ratio opposite in sign of the coefficients of the generalized Laguerre polynomial (1.80) of degrees N and $N - 1$:

$$S_1 = (4\tau)^{-1} N(2\tau - N + 1).$$

From these relations there immediately follows the simple expression for the energy E:

$$E = -\frac{Ng^2}{3}[(N - 1)(2N - 1) + 6\tau(\tau - N + 1)].$$

We shall now compute the frequencies of small oscillations nearby on equilibrium. Their squares are eigenvalues of the matrix

$$U_{jk} = \frac{\partial^2 H}{\partial x_j \partial x_k},$$

where $\{x_j\}$ are solutions to Equations (1.75a). Introducing the parameters $y_j = e^{2x_j}$ (like we did when calculating the equilibrium conditions), we write the matrix U as follows:

$$U_{jk} = 16g^2 \left[-(1 - \delta_{jk}(y_j^2 + y_k^2 + 4y_j y_k)(y_j - y_k)^{-4} \right.$$

$$+ \delta_{jk} \left(\sum_{m \neq j} (y_j^2 + y_k^2 + 4y_j y_k)(y_j - y_m)^{-4} \right.$$

$$\left. \left. + 2\tau^2 y_j (4y_j - 1) \right) \right]. \tag{1.81a}$$

Upon lenghty but not very difficult calculations, we find that under conditions (1.76) the matrix (1.81a) can be written in the form

$$U = Z^2, \quad Z_{jk} = 4g \left[-(1 - \delta_{jk}) \frac{y_j y_k}{(y_j - y_k)^2} \right.$$

$$\left. + \delta_{jk} \left(\sum_{m \neq j}^{N} \frac{y_j y_m}{(y_j - y_m)^2} + 2\tau y_j \right) \right]. \tag{1.81b}$$

Consequently, the frequencies we are looking for are eigenvalues of the matrix (1.81b) and are defined by the equations

$$(Z\lambda^{(n)})_j = 4g \sum_{k \neq j} \frac{y_j y_k}{(y_j - y_k)^2} (\lambda_j^{(n)} - \lambda_k^{(n)}) + 2\tau y_j \lambda_j^{(n)} = \omega_n \lambda_j^{(n)},$$

where $\lambda_j^{(n)}$ are components of the eigenvectors of the matrix Z corresponding to the eigenvalues $\omega_n, 1 \leq j, n \leq N$. These equations are similar to (1.76) in structure and their solutions can be investigated by the method used for relating the solutions to (1.76) with the generalized Laguerre polynomials. To this end, consider sums of the form

$$S_{jm} = \sum_{k \neq j}^{N} \frac{y_j y_k}{(y_j - y_k)^2} (y_j^{-m-1} - y_k^{-m-1} + 2\tau y_j^{-m}), \quad m > 0.$$

The integral over a contour encircling all the points $\{y_j\}$ and the origin of coordinates, of the function

$$\psi_j(y) = \frac{y y_j}{(y - y_j)^2} \left[y_j^{-m-1} - y^{-m-1} \frac{p_N'(y)}{p_N(y)} \right],$$

where $P_N(y)$ are above generalized Laguerre polynomials, is zero. Thus, the sum of residues of $\psi_j(y)$ at points $y = y_k, y = y_j$ and $y = 0$ is also equal to zero:

$$S_{jm} = -(\mathrm{res}\psi_j(y)|_{y=y_j} + \mathrm{res}\psi_j(y)|_{y=0}) + 2\tau y_j^{-m} \qquad (1.81c)$$

Let us calculate the first term of the right-hand side. At the point $y = y_j$, the function $\psi_j(y)$ has a third-order pole, and the residue at that point is given by the formula

$$\mathrm{res}\,\psi_j(y)|_{y=y_j} = \frac{d^2}{d\zeta^2} \left[\frac{y_j}{2}(\zeta + y_j)(y_j^{-m-1} \right.$$

$$\left. -(y_j + \zeta)^{-m-1} \frac{p_N'(y_j + \zeta)}{p_N(y_j + \zeta)} \right] |_{\zeta=0}.$$

The polynomial $p_N(y_j + \zeta)$ can easily be expanded in powers of ζ by using Equation (1.79) for p_N and expressions for its coefficients. The result is

$$\mathrm{res}\psi_j(y)|_{y=y_j} = -2\tau(m+1)y_j^{-m} + (m+1)\left(\frac{m}{2} + \tau - N + 1\right) y_j^{-m-1}.$$

The function $\psi_j(y)$ has a mth-order pole at the point $y = 0$, and the second term in the expression for S_m is a polynomial of degree m in the variable y_j^{-1},

$$\mathrm{res}\psi_j(y)|_{y=0} = \frac{y_j}{(m-1)!} \frac{d^{m-1}}{dy^{m-1}} \left[(y_j - y)^{-2} \frac{p_N'(y)}{p_N(y)} \right] |_{y=0}$$

$$= -\sum_{l=1}^{m-1} y_j^{-l+1} \frac{(l+1)}{(m-l-1)!} \frac{d^{m-l-1}}{dy^{m-l-1}} \frac{p_N'(y)}{p_N(y)} |_{y=0}.$$

Inserting all residues into (1.81c), we finally get

$$S_{jm} = \sum_{l=0}^{m} y_j^{-l+1} c_{cl}^{(m)},$$

$$c_m^{(m)} = (m+1)\left(\frac{m}{2} + \tau - N + 1\right), \qquad (1.81d)$$

Now, consider the action of the matrix Z (1.81b) on the N-dimensional vectors $\lambda^{(n)}$ with coordinates

$$\lambda_j^{(n)} = y_j^{-n}, \quad j = 1, \ldots, N.$$

The set of these vectors corresponding to the values of n from 1 to N forms a basis in the N-dimensional space since they are linear independent (but not orthogonal). In view of all these formulas, we have

$$(Z\lambda^{(n)})_j = 4g \sum_{l=1}^{N} c_{l-1}^{(n-1)} \lambda_j^{(l)}.$$

From this equality it follows that the matrix Z is triangular in the basis of $\{\lambda\}$, and its eigenvalues

$$\omega_n = 4g c_{n-1}^{(n-1)} = 4gn \left(\tau - N + \frac{n+1}{2} \right)$$

are determined by the formula (1.81d). The eigenvectors corresponding to these frequencies are linear combinations of the vectors $\{\lambda_j^{(n)}\}$. Their explicit form can be found by calculating the coefficients $c_{l-1}^{(n-1)}$, i.e. high-order derivatives of the logarithm of the generalized Laguerre polynomial at $y = 0$. The smallest frequency equals $\tau - N + 1$ and it becomes zero at the minimal possible value of τ. If τ is not a natural number, the frequencies are incommensurable, which proves the quasi-periodicity of finite motion in the general case.

So, we have found explicit expressions for the characteristics of equilibrium state that are commonly studied in the classical mechanics of the Sutherland systems in the Morse potential. It is seen that the existence conditions of equilibrium and the quantum discrete spectrum are equally dependent on the number of particles. What is more, as we shall show in the next chapter, at large values of the interaction constant g, i.e. $\hbar \ll 1$, the quantum and classical conditions coincide, as it should be according to quantum-mechanical laws.

1.6. Elliptic Solutions of Equations of Motion of Two Interacting Particles in an External Field

In the previous section, some hyperbolic solutions were found for classical Sutherland systems in the Morse potential. What happens if we consider the rational systems in an arbitrary quartic potential? The full answer will be presented in the next section, but it is of use to solve preliminarily the problem of two Calogero particles in the restricted quartic potential

$$V(x) = \frac{g^2}{x^2}, \tag{1.82a}$$

$$W(x) = \frac{1}{2}(Ax^2 + B)^2. \tag{1.82b}$$

We shall show that for arbitrary values of the parameters A, B and g^2 there exists a class of solutions of equations of motion of systems of that type containing two arbitrary constants and depending on time via the Jacobi elliptic functions; it is not of necessity to introduce the restriction $g^2 < 0$ indispensable for the method used in Section 1.4.

In what follows, we shall assume for convenience that $A > 0$ (without loss of generality by virtue of (1.82b)). The Hamiltonian of the considered system has the form

$$H = \frac{p_1^2 + p_2^2}{2} + \frac{g^2}{(x_1 - x_2)^2} + \frac{1}{2}[(Ax_1^2 + B)^2 + (Ax_2^2 + B)^2]. \tag{1.83}$$

Here, p_i and x_i are momenta and coordinates of particles. By using the Lax representation for the system (1.83), it is not difficult to write the second constant of motion as follows:

$$I^2 = |\xi_1 + i\xi_2|^2, \tag{1.84}$$

where

$$\xi_1 = p_1 p_2 - (Ax_1^2 + B)(Ax_2^2 + B) - \frac{g^2}{(x_1 - x_2)^2},$$

$$\xi_2 = p_1(Ax_2^2 + B) + p_2(Ax_1^2 + B).$$

From (1.84) it follows that

$$\frac{d}{dt}(\xi_1 + i\xi_2) = iS(\xi_1 + i\xi_2), \tag{1.85}$$

S being real. By direct calculation it can be easily verified that $S = 2A(x_1 + x_2)$; setting $xi_1 + i\xi_2 = Ie^{i\theta}$ we may express the coordinate of the center of mass of particles in terms of the derivative of the angle θ with respect to time:

$$x_1 + x_2 \frac{d}{dt}\theta/2A. \tag{1.86}$$

Using also the equation of motion

$$\frac{d}{dt}(p_1 + p_2) = -2A[x_1(Ax_1 + B) + x_2(Ax_2 + B)],$$

and the relation following from (1.83) and (1.84)

$$2(H + I\cos\theta) = \frac{d^2}{dt^2}{}^2 + \frac{1}{4}\frac{d}{dt}\theta^4(x_1 - x_2)^2 \tag{1.87}$$

we arrive at an equation for the angle θ only:

$$2\frac{d}{dt}\theta\frac{d^3\theta}{dt^3} - 3\frac{d^2\theta^2}{dt^2} + 4AB\frac{d\theta^2}{dt} + \frac{1}{4}\frac{d\theta^4}{dt}$$
$$+ 24A^2(H + I\cos\theta) = 0. \tag{1.88}$$

Note that all solutions of the equations of motion may be constructed from solutions of Equation (1.88), but the inverse statement is not valid. So, Equation (1.88) is not a result of the reduction on the basis of the known integrals (1.83) and (1.84). Nevertheless, as will be shown in what follows, with the help of (1.88) it is possible to find particular solutions to those equations corresponding to a certain choice of initial conditions.

We lower the order of Equation (1.88) by the substitution $\frac{d\theta}{dt}^2 = \nu(\theta)$, which gives the following equation for ν:

$$\nu\frac{d^2\nu}{d\theta^2} - \frac{3}{4}\left(\frac{d\nu}{d\theta}\right)^2 + 4AB\nu + \frac{1}{4}\nu^2 - 24A^2(H + I\cos\theta). \tag{1.89}$$

It may be verified from the right-hand side of (1.89) that the following solutions:

$$\nu(\theta) = 4A^2(x_1 + x_2)^2 = \frac{d^2\theta}{dt^2} = \lambda + \mu\cos\theta,$$

satisfy that equation under the condition that λ and μ obey the nonlinear system of equations

$$(\lambda - 8AB)\mu = 48A^2,$$

$$(\lambda + 8AB)^2 - 3\mu^2 = 64A^2B^2 - 96A^2H. \tag{1.90}$$

Thus, the time dependence of the coordinate of the center of mass of particles can be found by integrating equation (1.90) and then using (1.86). From (1.87) and (1.90), we may obtain also the relative coordinate as a function of the angle θ:

$$(x_1 - x_2)^2 = \frac{1}{4A^2}\left[-\frac{1}{3}(\lambda + 16AB) + \mu\cos\theta\right]. \tag{1.91}$$

However, as has been noted above, not every solution of (1.88) is associated with a trajectory of the considered system. This may be verified by substituting (1.91) and (1.86) into the equations of motion for the coordinate of the center of mass and the relative coordinate. The first of them holds for all values of the parameters λ and μ. The second is to be written in the form

$$u\frac{d^2u}{dt^2} - 1/2\left(\frac{du}{dt}\right)^2 = 128g^2A^4 - u^2\left[\frac{1}{4}(4\nu + u) + 4AB\right],$$

where the following notation is used:

$$\frac{1}{4A^2}u = (x_1 - x_2)^2,$$

$$\frac{1}{4A^2}\nu = (x_1 - x_2)^2 = \lambda + \mu\cos\theta.$$

Equation (1.93) only holds in the case when the constants λ and μ are related by the condition

$$\mu^2 = \frac{1}{9}(\lambda + 16AB)^2 - \frac{192g^2A^4}{\lambda + 4AB}. \tag{1.92}$$

Note that the sign of the parameter μ can always be chosen positive. Since $\mu^2 > 0$, not all values of the parameter λ are permissible. Condition (1.92) restricts the set of values of the constants of motion $\{H, I\}$ for which solutions to the equations of motion have the form (1.91). On the other hand, having constructed solutions in the form (1.91), we may parametrize them by one constant λ instead of I and H. Note also that the obtained trajectory satisfies the simple relation

$$x_1 x_2 = \frac{1}{12A^2}(\lambda + 4AB). \tag{1.93}$$

The second constant specifying the obtained solutions should be found by integrating the first-order equation as follows:

$$\left(\frac{d\theta}{dt}\right)^2 = \lambda + \mu \cos \theta. \tag{1.94}$$

Now, let us determine the interval of admissible values of λ. Apart from the inequality

$$\frac{1}{9}(\lambda + 16AB)^2 > \frac{192g^2 A^4}{\lambda + 4AB}$$

following from (1.92) we have the conditions for the right-hand side of (1.91) and (1.94) to be non-negative. For $B > 0$, this gives the following constraints on λ:

$$\lambda_1 < \lambda < -4AB, \tag{1.95}$$

where λ_1 is a negative root of the equation

$$(\lambda + 4AB)^2 (8AB - \lambda) = 216g^2 A^4.$$

This root does exist and does not exceed $-4AB$ for any values of the parameters $A, g^2, B > 0$, and, consequently, the set of admissible values of $\{\lambda\}$ is not empty (at $B = 0$ $\lambda_1 = 6(g^2 A^4)^{\frac{1}{3}}$). By substituting $y = \tan \frac{1}{2}\theta$, Equation (1.94) is reduced to the standard form

$$\left(\frac{dy}{dt}\right)^2 = \frac{1}{4}(\lambda + \mu)(1 + y^2)\left(1 + \frac{\lambda - \mu}{\lambda + \mu}y^2\right). \tag{1.96}$$

Values of λ from the interval (1.95) satisfy the condition $\mu > |\lambda|$, which allows us to write the solution to (1.96) as

$$\tan \frac{1}{2}\theta = \sqrt{\frac{\mu + \lambda}{\mu - \lambda}} \times \text{cn}\left(\sqrt{\frac{1}{2}(t - t_0)|\mu + \lambda|2\mu}\right), \tag{1.97}$$

where cn $(u|k^2)$ is the Jacobi elliptic cosine with modulus k. Together with condition (1.95) expression (1.97) completely determines, in accordance with (1.94) and (1.91), solutions to the equations of motion dependent on two parameters, λ and t_0.

When $B < 0$, the value for λ may also vary in the interval (1.95) that exists for any A, B and g^2. On the condition that

$$Ag^2/B < \frac{32}{243}B^2, \tag{1.98}$$

there may exist another type of motion. The parameter λ is also allowed to vary inside the interval $\tilde{\lambda}_1 < \tilde{\lambda}_2$, where $\tilde{\lambda}_1$ and $\tilde{\lambda}_2$ are positive roots of the equation

$$\frac{1}{9}(\lambda + 16AB)^2(\lambda + 4AB) = 192g^2A^4$$

not exceeding $-16AB$. According to (1.93), both particles are localized in one of two potential wells of the anharmonic oscillator $(Ax^2 + B)^2$; since in this case $\lambda > \mu$, the solution to equation (1.96) is of the form

$$\tan\frac{1}{2}\theta = \mathrm{sc}\sqrt{\frac{1}{2}(\lambda + \mu)}(t - t_0)|2\mu/(\lambda + \mu)),$$

where $\mathrm{sc}(u|k^2) = \mathrm{sn}(u|k^2)/\mathrm{cn}(u|k^2)$. From (1.98) it is seen that this type of motion occurs only in the case when the characteristic energy of repulsion of particles in the well (Ag^2/B) is significantly smaller than the height of the well $\frac{1}{2}B^2$.

Thus, we have found particular solutions of the equations of motion for a two-particle system with Hamiltonian (1.83) dependent on two arbitrary constants. These solutions exist for all values of the Hamiltonian parameters. In conclusion, we notice that a similar procedure allows also the determination of particular solutions for a more general class of Hamiltonians with two degrees of freedom disconnected with semisimple Lie algebras:

$$H = \frac{p_1^2 + p_2^2}{2} + \frac{g^2}{(x_1 - x_2)^2} + \frac{g^2}{(x_1 + x_2)^2} + \frac{1}{2}[(Ax_1^2 + B)^2 + (Ax_2^2 + B)^2]$$

The equations of motion for these Hamiltonians may be written in terms of the variables u and v as defined before Equation (1.92)

Integrable Many-Particle Systems

as follows:

$$v\frac{d^2v}{dt^2} - \frac{1}{2}\frac{dv^2}{dt} = 128g^2A^4 - v^2\left[\frac{1}{4}(3u+v) + 4AB\right],$$

$$u\frac{d^2u}{dt^2} - \frac{1}{2}\frac{du^2}{dt} = 128g^2A^4 - u^2\left[\frac{1}{4}(3v+u) + 4AB\right].$$

Setting here

$$u = \alpha_1 + \mu\cos\theta, \quad v = \alpha_2 + \mu\cos\theta,$$

$$\left(\frac{d\theta}{dt}\right)^2 = \lambda + \mu\cos\theta,$$

and performing the not difficult but rather lenghty computations, we verify that the last equation is a solution of the equations of motion provided that the parameter λ, μ, α_1 and α_2 are related by

$$\frac{3}{2}(\alpha_1 + \alpha_2) = \lambda - 8AB,$$

$$(\mu^2 - \alpha_1^2)(\alpha_1 - \lambda) = 256g^2A^4,$$

$$(\mu^2 - \alpha_2^2)(\alpha_2 - \lambda) = 256g^2A^4.$$

Together with positivity conditions for the expression (1.94), these relations allow one, like in the case of absence of the third term in the Hamiltonian, to establish the interval of feasible values of λ, one of the free parameters characterizing the solution to equations for u and v in the form (1.94). As the third term in the Hamiltonian is absent, these solutions pass over to the equations obtained above. The problem of explicit construction of the general solution to equations for u and v dependent on four arbitrary constants will be presented in the following sections.

1.7. Matrix Analogs of Elliptic Functions

In this section (which can be omitted at the first reading), we shall find the general solution of equations of motion for the rational Calogero system in the field defined by the polynomial of the fourth order without any restrictions on its parameters and initial condition. Moreover, we shall integrate some nonlinear matrix equation.

In 1980, Veselov proved the following beautiful result: Consider the Hamiltonian system

$$\frac{d^2 x_j}{dt^2} + \sum_{k \neq j} \frac{4}{(x_j - x_k)^3} + R(x_j) = 0, \quad j = 1, \ldots, N, \quad (1.99)$$

where $R(x)$ is an arbitrary polynomial. Then its solutions are eigenvalues of the matrix equation

$$\frac{d^2 Q}{dt^2} + R(Q) = 0 \quad (1.100)$$

with initial conditions

$$Q_{jk}(0) = x_j(0)\delta_{jk}, \quad \left(\frac{dQ}{dt}\right)_{jk}(0) = L_{jk}(0),$$

where $L_{jk} = \frac{dx_j}{dt}\delta_{jk} + 2(1 - \delta_{jk})(x_j - x_k)^{-1}$. For the polynomial $R(x)$ of the third order,

$$R(x) = a_0 x^3 + a_2 x + a_3, \quad (1.100a)$$

(terms of second order may be always removed by the shift of variables) the system (1.99) just gives the equations of motion defined by the Hamiltonian

$$H = \frac{1}{2}\sum_{j=1}^{N} p_j^2 + 2\sum_{j \neq k}(x_j - x_k)^{-2} + \sum_{j=1}^{N}\left(a_0 x_j^4/4 + a_2 x_j^2/2 + a_3 x_j\right),$$

which just coincides with our Hamiltonian (1.33). Hence, we shall study at first the matrix equation (1.100) with the polynomial (1.100a). Without loss of generality, we can put $a_0 = 2$. For the systems with the Hamiltonian

$$\tilde{H} = \frac{1}{2}\text{Tr}\left(\left(\frac{dQ}{dt}\right)^2 + Q^4 + a_2 Q^2 + 2a_3 Q\right),$$

and Poisson brackets

$$\left\{Q_{jk}, \left(\frac{dQ}{dt}\right)_{lm}\right\} P = \delta_{jm}\delta_{lk},$$

Equations (1.100) are equations of motion and have the Lax representation with the spectral parameter $h \in \mathbf{C}$:

$$\frac{dL}{dt} = [L, M], \quad L = \begin{pmatrix} \frac{dQ}{dt} & \psi_+(Q) \\ \psi_-(Q) & -\frac{dQ}{dt} \end{pmatrix},$$

$$M = \begin{pmatrix} 0 & \chi_+(Q) \\ -\chi_-(Q) & 0 \end{pmatrix},$$

$$\psi_\pm(Q) = Q^2 \pm ih\sqrt{2}Q + \left(\frac{a_2}{2} - h^2 \pm \frac{ia_3}{h\sqrt{2}}\right)E,$$

$$\chi_\pm(Q) = Q \pm \frac{iEh}{\sqrt{2}}, \tag{1.101}$$

where E is the unit matrix, $i^2 = -1$. The integrals of motion are defined uniquely by initial conditions as follows:

$$J_j = \mathrm{Tr}L^{2j}, \quad T_{jk} = \left[Q, \frac{dQ}{dt}\right]_{jk}, \quad 1 \le j, \ k \le N. \tag{1.102}$$

Let $\tilde{L}, \tilde{M}$ be matrices given from (1.101) by the similarity transformation with the matrix

$$Q_0 = \frac{1}{\sqrt{2}} \begin{pmatrix} E & E \\ -E & E \end{pmatrix}.$$

The elements with maximal degrees of h are now in their main diagonals, which allows one to integrate Equation (1.100) with the use of the methods of algebraic geometry developed by Dubrovin.

First, the spectral curve $z(h)$ is defined by the relation

$$S(z, h) = \det(\tilde{L}(h) - Iz) = 0,$$

where I is $2N \times 2N$ unit matrix. It gives covering Γ of the complex plane $\mathbf{C}$ with $2N$ sheets. The dimension of the Jacobian of this curve is defined by initial conditions. We consider first the case in which all eigenvalues τ_j of the commutator $\{T_{jk}\}$ are different.

Lemma 1. *The genus g of the curve $z(h)$ equals $N^2 - N + 1$ for $a_3 = 0$ and $N^2 + 1$ for all other values of a_3.*

Proof. In the neighborhood of $h = 0$ the branches of z have asymptotic

$$z_j = a_3/h\sqrt{2} + \nu_j h + O(h), \quad z_{j+N} = -a_3/h\sqrt{2} + \nu_j h + O(h),$$

where ν_j are eigenvalues of the isospectral matrix R,

$$R = i\left[\frac{dQ}{dt}, Q^2\right] + Q'^2 + Q^4 + a_2 Q^2 + 2a_3 Q + a_2^2\frac{E}{4}.$$

The function $S' = \frac{\partial S}{\partial z}$ has on Γ at $a_3 \neq 0$ simple poles in the images of the point $h = 0$. In the neighborhood of the points $P_{j1}^{(0)}, P_{j2}^{(0)}, 1 \leq j \leq N$ which are images of $h^{-1} = 0$, for the branches of $z(h)$ the following decompositions take place:

$$z_{j\alpha} = (-1)^{\alpha+1}\left(h^2 - \frac{a_2^2}{2}\right) + i\tau_j/h\sqrt{2} + o(h^{-1}), \quad \alpha = 1, 2;$$

S' has on each sheet in these points the poles of the degree $N + 1$. The total number of the poles of S' on Γ is the number of the branch points of Γ, and the proof of the Lemma 1 follows from the Riemann–Gurwitz formula.

Let us introduce, following Dubrovin, the notion of multidimensional theta function of the Riemann surface Γ. Let $B = \{B_{jk}\}$ be the symmetric $g \times g$ matrix with a negatively defined real part. It is often called Riemann matrix. θ-function of Γ is given by the g-dimensional Fourier series

$$\theta_\Gamma(x) = \sum_{n \in \mathbf{Z}^g} \exp\left[\frac{1}{2}(Bn, n) + (n, x)\right],$$

where x is a vector with g components and elements of the matrix B are calculated by integration of some differentials over some contours on Γ (a more detailed definition can be found in the review by Dubrovin).

Let ω be the matrix of transformation which diagonalizes the commutator (1.102), $(\omega T\omega^{-1})_{jk} = \tau_j \delta_{jk}$; $\tilde{L}_1, \tilde{M}_1$ are matrices which are obtained from $\tilde{L}, \tilde{M}$ by the similarity transformation generated by the matrix

$$\Omega_1 = \begin{pmatrix} \omega & 0 \\ 0 & \omega \end{pmatrix}; \quad P \in \Gamma.$$

The eigenvectors of the matrix $\tilde{L}_1$ which satisfy the equation

$$\left(\frac{d}{dt} + \tilde{M}_1 \right) \lambda(t, P) = 0$$

have essential singularities in the points $P_{j\alpha}^{(0)}$ of the type $\exp(\pm iht/\sqrt{2})$. Let us numerate the sheets of Γ and coordinates of vectors $\lambda(t, P)$ by two indices so that $P_{j\alpha}^{(0)}$ can be on the sheets $(j\alpha)$ and

$$(\lambda(t, P))_{k1} = (\lambda(t, P))_k, \quad (\lambda(t, P))_{k2} = (\lambda(t, P))_{k+N}, \ 1 \le k \le N.$$

Let Λ be the matrix of which vectors $\lambda(t, P_{j\alpha})$ are columns, vectors $\mu(t, P_{j\alpha})$ are the rows of Λ^{-1}, $D^{j\alpha}(t), D_{j\alpha}(t)$ are divisors of zeros of $(\lambda(t, P))_{j\alpha}, (\mu(t, P))_{j\alpha}$.

Lemma 2. *The degrees of divisors $D(t)$ are equal to g.*

The proof consists in construction of the asymptotic of the projector

$$g_{j\alpha,k\beta}(t, P) = (\lambda(t, P))_{j\alpha}(\mu(t, P))_{k,\beta},$$

at $P \to P_{i\gamma}^{(0)}$ with the use of the diagonality of T in the chosen representation for $\tilde{L}_1, \tilde{M}_1$ and use the results of Lemmas 5, 6 in Dubrovin's work.

Let us define the functions $\tilde{f}(t, P)$ for the functions $f(t, P)$ defined on Γ, by the relation

$$\tilde{f}(t, P) = \frac{f(t, P)}{f(0, P)}.$$

The dependence of Q on time is defined by the values of $\tilde{g}_{j1,k2}(t, P)$ at the points $P_{j1}^{(0)}$.

Proposition 1. *The functions $(\tilde{\lambda}(t, P))_{j\alpha}, (\tilde{\mu}(t.P))_{j\alpha}$ are meromorphic on $\Gamma \backslash (P_{j\gamma}^{(0)})$, divisors of their poly have degree g and are defined* *by initial conditions for matrices* Q. *In the neighborhoods of the points* $P_{j\gamma}^{(0)}$, *these functions have the form*

$$\chi(t, j\alpha, l\gamma) \exp[\pm(-1)^{\gamma+1} i h t \sqrt{2}],$$

i.e. they are $2N$-*point Baker–Akhiezer functions and may be expressed via* θ-*functions of the Riemann surface* Γ.

Let $\{a_p, b_q\}, 1 \leq p, q \leq g$-canonical basis of closed contours on Γ; $\{\omega_p\}$ is normalized basis of the differentials of the first kind; Ω is normalized differential of the second kind with main parts of the type $i(-1)^{\gamma+1} dh/\sqrt{2}$ in the points $P_{l\gamma}^{(0)}$; K is the vector of Riemann constants with coordinates

$$K_p = \pi i + \oint_{b_p} \omega_p - \frac{1}{2\pi i} \sum_{q \neq p} \oint_{a_q} \left(\omega_q(P) \int_{P_0}^{P} \omega_p \right),$$

$A(P)$ is the Abel transformation. The vector $W = K + A(P_{j1}(0)) - A(P_{j2}(0))$ on the Jacobian $J(\Gamma)$ does not depend on j. The solution of Equation (1.100) with (1.101b) has the form $Q_{jk}(t) = (\omega^{-1}\tilde{Q}\omega)_{jk}$, where

$$\tilde{Q}_{jk}(t) = (\omega Q(0)\omega^{-1})_{jk} \exp(t(\xi_j - \tilde{\xi}_k))$$

$$\times \frac{\theta_\Gamma(A(P_{k2}^{(0)}) - A(P_{j1}^{(0)}) + tU + W)\theta_\Gamma(W)}{\theta_\Gamma(A(P_{k2}^{(0)}) - A(P_{j1}^{(0)}) + W)\theta_\Gamma(tU + W)}, \quad (1.102)$$

$$\xi_j = \lim_{P \to P_{j1}^{(0)}} \left(\int_{P_0}^{P} \Omega - \frac{ih}{\sqrt{2}} \right), \quad \tilde{\xi}_k = \lim_{P \to P_{k2}^{(0)}} \left(\int_{P_0}^{P} \Omega + \frac{ih}{\sqrt{2}} \right),$$

$$(U)_P = \oint_{b_p} \Omega,$$

where θ_Γ are Riemann theta functions of the surface Γ. From the structure of L-matrix it follows the existence for Γ involution ρ : $(z \to -z, h \to -h)$ and existence of two-sheet covering π: $S(z, h) = 0 \to S_0(\tilde{z}, \tilde{h}) = 0$, where $\tilde{z} = zh, \tilde{h} = h^2$. In the case of $a_3 \neq 0$, branch points of π coincide with the images of $h = 0$, for $a_3 = 0$, the covering π has no branch points. In both cases the genus of the curve S_0, which is calculated according to Riemann–Gurwitz formula, equals

$$g_0 = \frac{1}{2}(N^2 - N + 2).$$

Since τ interchanges the sheets of Γ and singular points of the Baker–Akhiezer functions $\tilde{\lambda}(t, P), \tilde{\mu}(t, P)$, the dynamics is realized not on the corresponding to ρ Prym variety but on the Jacobian $J(S_0)$, and theta functions are reduced to the dimension g_0.

For the particle systems (1.99), the commutator T (1.102) has the form $T_{jk} = 2i(1 - \delta_{jk})$, its eigenvalue $-2i$ is $(N - 1)$-times degenerated. The genus of the curve $S(z, h)$ is lowering to $2N - 1$ at $a_3 = 0$ and $3N - 1$ for other cases; the dimension of the Jacobian $J(S_0)$ equals N in both cases.

So, we find the solution of equations of motion of the Calogero particle systems in quartic external potential generated by a particle of infinite mass. Its complexity differs drastically from the solutions of equations of motion for Sutherland systems in the Morse potential. To our best knowledge, nobody was able to solve the equations of motion for the Sutherland systems in the general hyperbolic potential (1.33).

1.8. Extension of the Class of Integrable Dynamical Systems Connected with Semisimple Lie Algebras

Here, we shall consider more general dynamical systems with the Hamiltonian

$$H = \sum_{j=1}^{N} \frac{p_j}{2} + \sum_{j>k}^{N} g^2 [V(x_j - x_k) + V(x_j + x_k)] + \sum_{j=1}^{N} W(x_j). \quad (1.103)$$

It corresponds to the motion of symmetric ensemble of $2N$ particles which are posed symmetrically near the origin and have opposite momenta. For the special case $W(x) = g_1^2 V(x) + g_2^2 V(2x)$, the Hamiltonian (1.103) has the structure related with the systems of roots of semisimple Lie algebras B_N, C_N, BC_N. This case was discovered and intensively studied by Olshanetsky and Perelomov.

Our goal consists in finding new potentials $W(x)$ for which the systems described by (1.103) are integrable.

The main result of Olshanetsky and Perelomov consists in the following: the Lax matrices exist and integrability takes place if

$$g_1^2 + \sqrt{2}gg_2 - 2g^2 = 0,$$

or $g_1 = 0, g, g_2$ are arbitrary and

$$V(x) = \left\{ \frac{1}{x^2}, \frac{1}{\sin\mathrm{h}^2 ax}, \wp(ax) \right\},$$

where $\wp$ is the Weierstrass function. Contrary to their hypothesis, we shall use for the Lax matrices L, M the following generalized Ansatz:

$$L = \begin{pmatrix} l & \lambda & \psi \\ \lambda^+ & 0 & -\lambda^+ \\ \psi^+ & -\lambda & -l \end{pmatrix},$$

$$M = \begin{pmatrix} m & \omega & s \\ -\omega^+ & \mu & -\omega^+ \\ -s^+ & \omega & m \end{pmatrix},$$

where l, ψ, m, s are $N \times N$ matrices, and λ, ω are $N \times 1$ matrices dependent on dynamic variables (p_j, x_j):

$$l_{jn} = p_j \delta_{jn} + i(1 - \delta_{jn})gR(x_j - x_n),$$

$$\psi_{jn} = i[\delta_{jn}(\nu(x_j) + i\rho(x_j)) + (1 - \delta_{jn})gR(x_j + x_n)],$$

$$s_{jn} = i\left[\frac{\delta_{jn}}{2}(\nu'(x_j) + i\rho'(x_j)) + (1 - \delta_{jn})gR'(x_j - x_n) \right],$$

$$m_{jn} = i\delta_{jn}\left[\tau(x_j) - \sum_{k \neq j} g(z(x_j - x_n) + z(x_j + x_n)) \right] \qquad (1.104)$$

$$+ i(1 - \delta_{jn})gR'(x_j - x_n),$$

$$\lambda_j = i\alpha(x_j), \quad \omega_j = i\alpha'(x_j), \quad \mu = i\sum_{j=1}^{N} \kappa(x_j).$$

Note that L is an operator of an irreducible matrix representation of the Lie algebra corresponding to the $SU(N+1, N)$ group.

From the relation $H = \frac{1}{2}\operatorname{Tr}(L^2)$, it follows that the function $V(x)$ and $W(x)$ determining potential (1.103) are connected with $R(x)$, $\nu(x)$, $\rho(x)$ and $\alpha(x)$:

$$V(x) = g^2 R^2(x), \quad W(x) = \frac{1}{2}(\nu^2(x) + \rho^2(x) + \alpha^2(x)).$$

The Lax equation imposes constraints on the functions $R, \tau, \nu, \rho, z, \alpha$ and κ in (1.104). First, R and z should obey the equation

$$R'(x)R(y) - R'(y)R(x) = R(x+y)(z(x) - z(y))$$

solutions to which have been found by Calogero and Moser as follows:

$$R(x) = \{x^{-1}, (\sinh ax)^{-1}, (\operatorname{sn}(ax, k))^{-1}\},$$
$$z(x) = \{-x^{-2}, -a(\sinh ax)^{-2} - a(\operatorname{sn}(ax, k))^{-2}\},$$

where $\operatorname{sn}(ax, k)$ is the Jacobi elliptic function with modulus k. Second, for other functions from (1.104), we obtain two sets of functional equations, and in each case some of the functions α, k and ρ vanish:

$$(I) \quad \alpha(x) = \kappa(x) = 0,$$

$$2R(x + \varepsilon y)(\tau(x) - \tau(y)) - R(x - \varepsilon y)(\nu'(x) + \nu'(y))$$
$$-2R'(x - \varepsilon y)(\nu(x) - \varepsilon\nu(y)) = 0, \qquad (1.105a)$$
$$R(x + \varepsilon y)(\rho'(x) - \varepsilon\rho'(y)) + 2R'(x + \varepsilon y)(\rho(x) - \rho(y)) = 0,$$

$$\varepsilon = \pm 1.$$

$$(II) \quad \rho(x) = 0.$$
$$2R(x + \varepsilon y)(\tau(x) - \tau(y)) - R(x - \varepsilon y)(\nu'(x) + \nu'(y))$$
$$-2R'(x - \varepsilon y)(\nu(x) - \varepsilon\nu(y))$$
$$= 2(\alpha(x)\alpha'(y) - \varepsilon\alpha'(x)\alpha(y))g^{-1},$$

$$[R(x-y) + R(x+y)]\alpha'(y) + [R'(x-y) - R'(x+y)]\alpha(y)$$
$$+ [g(z(x-y) + z(x+y)) + \kappa(y)]\alpha(x) = 0,$$
$$\alpha'(x)\nu(y) = \left[\tau(x) - \frac{\nu'(x)}{2} - \kappa(x)\right]\alpha(x). \qquad (1.105b)$$

The method of finding analytic solutions to these equations consists of expanding the latter in series about point $y = 0$ with due consideration to possible pole singularities of the functions to be found and in solving the resulting ordinary differential equations. We shall present here only the final results.

(I) From the expressions for $R(x)$ and $z(x)$, it follows that

$$\nu(x) = [\alpha_1 + \beta_1 \operatorname{sn}^2(ax, k) + \gamma_1 \operatorname{sn}^4(ax, k)]$$
$$\times [\operatorname{sn}(ax, k)\operatorname{cn}(ax, k)\operatorname{dn}(ax, k)]^{-1},$$
$$\tau(x) = a\nu(x)[\operatorname{sn}(2ax, k)]^{-1}.$$

at $R(x) = (\operatorname{sn}(ax, k))^{-1}$; α_1, β_1 and γ_1 are arbitrary constants. For other values of R which are limiting cases of $(\operatorname{sn}(ax, k))^{-1}$, the functions $\nu(x)$ and $\tau(x)$ can be obtained by taking the corresponding limit.

All solutions to (1.105a) at $\varepsilon = -1$ have been found by Inozemtsev. For $\varepsilon = 1$, Equations (1.105a) have only the following solutions from the above class:

$$\rho(x) = \alpha_2 + \gamma_2 x^2, \quad R(x) = x^{-1},$$
$$\rho(x) = \alpha_2 + \gamma_2 \cosh(2ax), \quad R(x) = (\sinh ax)^{-1},$$

where α_2 and γ_2 are arbitrary; at $R(x) = (\operatorname{sn}(ax, k))^{-1}$, non-trivial solutions of (1.105a) are absent.

According to these equations, in the considered case we arrive at the following sets of functions $\{V, W\}$:

$$V(x) = g^2 x^{-2}, \quad W(x) = g_1^2 x^{-2} + g_2^2 x^2 + g_3^2 x^4 + g_4 x^6,$$
$$V(x) = g^2 (\sinh(ax/2))^{-2},$$

$$W(x) = g_1^2(\sinh(ax/2))^{-2} + g_2^2(\sinh ax)^{-2}$$
$$+ g_3^2 \cosh ax + g_4^2 \cosh(2ax), \tag{1.106}$$

where all the constants $g_\gamma (\gamma = 1, \ldots, 4)$ are arbitrary. Using the relation

$$\wp(ax) = \frac{1}{\operatorname{sn}^2(ax, k)},$$

where $\wp$ is the Weierstrass function with two periods ω_1, ω_2 defined by the modulus k of the Jacobi elliptic functions, we obtain the third solution for (V, W):

$$V(x) = g^2 \wp(ax),$$
$$W(x) = g_1^2 \wp(ax) + g_2^2 \wp\left(ax + \frac{\omega_1}{2}\right) + g_3^2 \wp\left(ax + \frac{\omega_2}{2}\right)$$
$$+ g_4^2 \wp\left(ax + \frac{\omega_1 + \omega_2}{2}\right). \tag{1.107}$$

Constants g_γ here are not arbitrary but should satisfy the nonlinear equation

$$\left(\sum_{\gamma=1}^{4} g_\gamma^4 - \sum_{\beta>\gamma}^{4} g_\beta^2 g_\gamma^2\right)^2 = 64 \prod_{\gamma=1}^{4} g_\gamma^2.$$

(II) In this case, we determine α and κ from the Equation (1.105b). For $R(x) = (\operatorname{sn}(ax, k))^{-1}$, we have

$$\alpha(x) = a_1(\operatorname{sn}(ax, k))^{-1},$$
$$\kappa(x) = 2ga(\operatorname{sn}(ax, k))^{-2}.$$

Substituting $\tau(x) = \tau_1(x) + \tau_2(x)$ with $\tau_1(x) = aa_1 g^{-1}(\operatorname{sn}(ax, k))^{-2}$, we may transform the system of equation (1.105b) to equation (1.105a) and make use of the known solution of (1.105a) to determine $\nu(x)$ and $\tau_2(x)$. The last of equation (1.105b) restricts the choice of constants in the functions $\tau_2(x)$, $\alpha(x)$, $\kappa(x)$ and $\nu(x)$ thus obtained.

The general solution to equation (1.105b) can be written as follows:

$$\alpha(x) = a_1(\mathrm{sn}(ax,k))^{-1},$$

$$\kappa(x) = 2ga(sn(ax,k))^{-2},$$

$$\nu(x) = [\alpha_1 + \gamma_1 \, \mathrm{sn}^4(ax,k)][\mathrm{sn}(ax,k)\mathrm{cn}(ax,k)\mathrm{dn}(ax,k)]^{-1},$$

$$\tau(x) = aa_1 g^{-1}(\mathrm{sn}(ax,k))^{-2} + a\nu(x)(\mathrm{sn}(2ax,k))^{-1},$$

$$a_1^2 = 2g^2 - 2\alpha_1 g,$$

where α_1 and γ_1 are arbitrary constants. The functions V, W corresponding to these relations have the form (1.107) and the admissible points in the four-dimensional space $\{g_\gamma\}$ are in a two-dimensional hypersurface defined by the equations

$$F(k^4 g_2^2, (1+k^2)^2 g_3^2, g_4^2) = 0,$$

$$F\left(g_1^2, g_3^2 + \frac{1-k^2}{1+k^2}(g_2^2 - g_4^2), 2g^2\right) = 0, \qquad (1.108)$$

where

$$F(x,y,z) = x^2 + y^2 + z^2 - 2xy - 2xz - 2yz \qquad (1.109)$$

or by equations following from the latter through the interchanges:

$$\{g_1 \leftrightarrow g_2, g_3 \leftrightarrow g_4\},$$

$$\{g_1 \leftrightarrow g_3, g_2 \leftrightarrow g_4\},$$

$$\{g_1 \leftrightarrow g_4, g_2 \leftrightarrow g_3\}.$$

These interchanges correspond to a shift of all coordinates by half-periods of $\wp(ax)$.

Thus, we have established the existence of the Lax matrices for Hamiltonian systems with potential (1.107) defined by the set of functions (1.105b) and the elliptic potential with restriction for the coupling constants (1.108) and (1.109).

All these systems possess the additional integrals of motion $I_n = 1/4N \, \mathrm{Tr}(L^{2n})$, $n = 1, \ldots, N$ which are in involution. The last statement can be proved by using the method of Section 1.3, by a direct calculation of the Poisson brackets of any two eigenvalues of the matrix L.

All the previously known integrable systems described in the previous sections can be obtained from the systems (1.107) with the potential (1.103) by some limiting transitions including shifts of coordinates and "renormalization" of coupling constants $\{g_\gamma\}$. In particular, the results of Olshanetsky and Perelomov correspond to particular solutions of (1.108) of the form

$$g_1^2 = g_2^2 = g_3^2 = g_4^2,$$
$$g_2^2 = g_3^2 = g_4^2, \quad g_1^2 = 2g^2 - \sqrt{2}gg_2.$$

This can easily be established by using the relation

$$\wp(2ax) = \frac{1}{4}\left[\wp(ax) + \wp\left(ax + \frac{\omega_1}{2}\right) + \wp\left(ax + \frac{\omega_2}{2}\right)\right.$$
$$\left. + \wp\left(ax + \frac{\omega_1 + \omega_2}{2}\right)\right].$$

It is natural to ask whether it is possible to find the elliptic systems of the type (1.106) without any restrictions to coupling constants $\{g\}$. This question will be considered in the last section of this chapter.

1.9. The Finite Toda Lattices

The study of the propagation of waves in infinite one-dimensional lattices, as it was shown first by Toda, made it possible to derive exact analytic solutions to an infinite system of nonlinear differential equations describing the exponential interaction between the nearest neighboring particles. Once Henon, Flashka and Moser showed complete integrability of finite-dimensional systems of that type (non-periodic and periodic Toda lattices), a number of papers appeared devoted to the investigation of their properties. Kostant, Olshanetsky and Perelomov established the connection between non-periodic lattices and classical Lie algebras. Hamiltonians of those lattices may be constructed with the use of simple roots $\{r\}$ of classical Lie algebras $\mathcal{G}$,

$$H = \sum_{j=1}^{N} \frac{p_j^2}{2} + V_{\mathcal{G}}, \quad V_{\mathcal{G}} = \sum_{\alpha \in \{r\}} \exp(\alpha q),$$

where α are root vectors, p_j and q_j are respectively momenta and coordinates of particles. For the algebras A_{N-1}, B_N, C_N and D_N, the potentials $V_{\mathcal{G}}$ are of the form

$$V_{A_{N-1}} = \sum_{j=1}^{N-1} \exp(q_j - q_{j+1}), \quad V_{B_N} = V_{A_{N-1}} + \exp(q_N),$$

$$V_{C_N} = V_{A_{N-1}} + \exp(2q_N), \quad V_{D_N} = V_{A_{N-1}} + \exp(q_{N-1} + q_N).$$
$$\tag{1.110a}$$

For the exceptional algebra systems like (1.110a), no simple mechanical interpretations are present and none will be considered here. Bogoyavlensky has pointed out a method for constructing generalized non-periodic lattices that, as first shown in the papers of Reyman and Semenov–Tjan–Shansky, correspond to the use of a system of simple roots of affine algebras. Potentials V for loop algebras $A_N^{(1)}$, $B_N^{(1)}$, $C_N^{(1)}$ and $D_N^{(1)}$ can be written in the form

$$V_{A_N^{(1)}} = V_{A_{N-1}} + \exp(q_N - q_1),$$

$$V_{B_N^{(1)}} = V_{B_N} + \exp(-q_1 - q_2),$$

$$V_{C_N^{(1)}} = V_{C_N} + \exp(-2q_1),$$
$$\tag{1.110b}$$

$$V_{D_N^{(1)}} = V_{D_N} + \exp(-q_1 - q_2).$$

As for the twisted loop algebras, they correspond to the potentials found by Olshanetsky, Perelomov, Reyman and Semenov–Tjan–Shansky:

$$V_{A_{2N}}^{(2)} = V_{A_{N-1}} + \exp(q_N) + \exp(-2q_1),$$

$$V_{A_{2N+1}}^{(2)} = V_{A_{N-1}} + \exp(-q_1 - q_2) + \exp(2q_N),$$
$$\tag{1.110c}$$

$$V_{D_{N+1}}^{(2)} = V_{A_{N-1}} + \exp(q_N) + \exp(-q_1).$$

The Lax representation for generalized periodic lattices (1.110b) and (1.110c) depend on spectral parameter, and the corresponding Hamiltonian flows are linearized on the Jacobi varieties of the algebraic curves associated with the L-matrix spectrum as it was found by Adler and Moerbeke. The list (1.110a)–(1.110c) covers all the potentials corresponding to infinite series of the Kac–Moody algebras. The existence of other, different from (1.110a)–(1.110c), integrable systems with exponential potentials like $V = \sum_{j=1}^{N+1} \exp(\sum_{k=1}^{N} P_{jk} q_k)$ also have been questioned by Adler and van Moerbeke. They have shown that when the rank of P equals N, the Kowalewski–Painleve property pertains only to the trajectories of systems (1.110a)–(1.110c). Matrices of a smaller rank were not analyzed.

We shall show that all the above-listed Toda lattices are particular cases of the systems described in Section 1.8 corresponding to certain limit situations. And what is more, taking advantage of that correspondence, we obtain new integrable lattices. This does not contradict with the results of Adler and Moerbeke as they may be interpreted as systems with matrices P_{jk} of a rank smaller than the number of degrees of freedom. Also, lattices will be found for which the interaction of particles at the ends is not exponential.

Let us now show how from (1.103) and (1.107) one can obtain potentials of the Toda lattices (1.110a)–(1.110c). Let us write (1.103) in the form

$$H = \sum_{j=1} p_j^2/2 + U_1(q) + \varepsilon(U_2(q) + U_3(q)), \quad \varepsilon = 0, 1,$$

where

$$U_1(q) = g^2 \sum_{j>k}^{N} \sinh^{-2}\left(\frac{q_j - q_k}{2}\right),$$

$$U_2(q) = g^2 \sum_{j>k}^{N} \sinh^{-2}\left(\frac{q_j + q_k}{2}\right),$$

$$U_3(q) = \sum_{j=1}^{N} \left[\frac{g_1^2}{\sinh^2 q_j/2} + \frac{g_2^2}{\sinh^2 q_j} + g_3^2 \cosh q_j + g_4^2 \cosh 2q_j\right],$$

or

$$U_1(q) = g^2 \sum_{j>k}^{N} \wp(q_j - q_k),$$

$$U_2(q) = g^2 \sum_{j>k}^{N} \wp(q_j + q_k),$$

$$U_3(q) = \sum_{j=1}^{N} [g_1^2 \wp(q_j) + g_2^2 \wp(q_j + \omega_1/2)$$

$$+ g_3^2 \wp(q_j + \omega_2/2) + g_4^2 \wp(q_j + (\omega_1 + \omega_2)/2)].$$

for elliptic potential. Note at first that all the constants g, $g_1 - g_4$ in (1.103) are arbitrary. The same concerns, as it will be shown in the next section, the constants in elliptic potential. Introduce now variables x_j by the relations

$$q_j = x_j + (j-1)\Delta, \quad \Delta > 0. \tag{1.111}$$

Then, we set

$$g = \frac{g_0}{2} \exp(\Delta/2),$$

and take the limit $\Delta \to \infty$ with q_j as in (1.111). As at large Δ one may expand $\left[\sinh\left(\frac{q_j - q_k}{2}\right)\right]$,

$$\left[\sinh\left(\frac{q_j - q_k}{2}\right)\right] \sim 4\exp(-\Delta(j-k))(\exp(x_k - x_j) + O(e^{-\Delta(j-k)}),$$

$$j > k$$

in the potential $U_1(q)$ when $\Delta \to \infty$ there remain only terms corresponding to the interaction of nearest-neighbor particles:

$$\tilde{U}_1(x) = \lim_{\Delta \to \infty} U_1(q) = g_0 V_{A_{N-1}}(x).$$

Thus, at $\varepsilon = 0$ we obtained the potential of non-periodic Toda lattice from the Calogero–Moser potential. Let us clarify the matter with other terms. It is easy to see that U_2 now contains only one term,

$$\tilde{U}_2(x) = \lim_{\Delta \to \infty} U_2(q) = g_0 \exp(-x_1 - x_2).$$

Inserting (2.93) into the term $U_3(q)$ and using the decompositon of $\sinh q$ at large Δ, we observe that the limit when $\Delta \to \infty$ is finite only when g_1 and g_2 do not depend on Δ, whereas g_3 and g_4 are exponentially decreasing, $g_3^2 \sim Ce^{-\Delta(N-1)}$ and $g_4^2 \sim De^{-2\Delta(N-1)}$,

$$\tilde{U}_3(x) = \lim_{\Delta \to \infty} U_3(q) = g_0 \left[\frac{A}{\sinh^2 x_1/2} + \frac{B}{\sinh^2 x_1} + Ce^{x_N} + De^{2x_N} \right].$$

Here, the constants A, B, C, D are arbitrary. Thus, we have got the system with the potential

$$V_1(x) = \tilde{U}_1(x) + \varepsilon[\tilde{U}_2(x) + \tilde{U}_3(x)] = g_0 V_{A_{N-1}}(x)$$

$$+ \varepsilon g_0 \left[e^{-x_1-x_2} + \frac{A}{\sinh^2 x_1/2} + \frac{B}{\sinh^2 x_1} + Ce^{x_N} + De^{2x_N} \right].$$
$$(1.112)$$

When $A = B = 0$ and $\varepsilon = 1$, this represents the potential of the lattice with exponential interaction

$$V_2(x) = g_0[V_{A_{N-1}}(x) + e^{-x_1-x_2} + Ce^{x_N} + De^{2x_N}]. \quad (1.113)$$

The cases $D_N, B_N^{(1)}$ and $A_{2N+1}^{(2)}$ follow from (1.113) with the choice of constants $C = D = 0$, $D = 0$ and $C = 0$, respectively. Any of the non-zero constants in (1.113) can be made equal to 1 by a finite shift of all the coordinates. Shifting the whole system to the right, $x_j \to x_j + \Delta_1$, $\Delta_1 > 0$, $1 \le j \le N$, and performing "renormalization" of the constants

$$\tilde{A} \to \frac{Ae^{\Delta_1}}{4}, \quad \tilde{B} \to \frac{Be^{2\Delta_1}}{4},$$
$$C \to Ce^{-\Delta_1}, \quad D \to De^{-2\Delta_1},$$

we obtain from (1.110), when $\Delta_1 \to \infty$, the potential

$$V_3(x) = g_0[V_{A_{N-1}}(x) + Ae^{-x_1} + Be^{-2x_1} + Ce^{x_N} + De^{2x_N}]. \quad (1.113a)$$

This potential has been independently found by Sklyanin via Yang–Baxter equation technique. It describes the motion of the Toda system whose extreme particles interact with an external field, the interaction potentials being different for the first and last particles. When part of the constants vanish, we obtain the known lattices

$$V_{B_N} : \ C = 1, A = B = D = 0, \quad V_{A_{2N}^{(2)}} : \ C = B = 1, A = D = 0,$$

$$V_{C_N} : \ D = 1, A = B = C = 0, \quad V_{D_{N+1}^{(1)}} : \ A = C = 1, B = D = 0,$$

$$V_{C_N^{(1)}} : \ B = D = 1, A = C = 0.$$

By finite shifts of the type $x_j \to x_j + \delta_1 + j\delta_2$ and change of the time scale, one of the non-zero constants can be made equal to 1 in each of the pairs (A, B), (C, D) so that the potential actually contains only two arbitrary constants.

Thus, from the trigonometric degenerations of the general potentials with four Weierstrass functions obtained in the previous section we have obtained most parts of the known lattices with exponential interactions (1.110a)–(1.110c) and some new ones. There are, however, two cases: $A_N^{(1)}$ (the periodic lattice) and $D_N^{(1)}$ that cannot be derived from trigonometric potentials in any limit. Nevertheless, they may also be included in the general elliptic potential. To demonstrate this, note that we passed from elliptic potential to (2.94) via two stages: first we tended the real-valued period of the Weierstrass function, ω_2, to infinity, and then carried out shifts of the coordinates of particles without correlation with ω_2. Now, let us put

$$q_j = x_j + (j - 1)\omega_2\tau, \quad \tau > 0 \tag{1.114}$$

and use for $\wp(\zeta)$ the representation ($\omega_1 = 2\pi i$):

$$\wp(\zeta) = \frac{1}{4} \sum_{m=-\infty}^{\infty} \left[\sinh\left(\frac{\zeta - m\omega_2}{2} \right) \right]^{-2} + C,$$

$$C = \frac{1}{12} - \frac{1}{2} \sum_{m=1}^{\infty} \frac{1}{\sinh^2(m\omega_2/2)}.$$

In the limit of large values of the argument and real-valued period, we derived from previous equation the expansion

$$\wp(\zeta + \omega_2 \delta) = C + \exp\left[-\zeta(\delta - 1/2)|\delta - 1/2| - 1\right.$$
$$\left. - \omega_2(1/2 - |\delta - 1/2|)\right] + O(e^{-\omega_2(1/2 - |\delta - 1/2|)}),$$

where $0 < \delta < 1, \delta \neq 1/2$.

We put $g^2 = e^{\omega_2 \tau}$ and calculate the limits of the elliptic potentials $U_1(q)$ and $U_2(q)$ with the coordinates (1.113) when $\omega_2 \to \infty$. From the previous equation it follows that the limit of $U_1(q)$ may contain an additional to the usual $V_{A_{N-1}}(x)$ term only for the choice $\tau = \frac{1}{N}$ when there are valid expansions

$$\wp(q_j - q_{j+1}) \sim C + \exp(x_j - x_{j+1}) \times \exp(-\omega_2/N),$$
$$\wp(q_1 - q_N) \sim C + \exp(x_N - x_1) \exp(-\omega_2/N),$$
$$\lim_{\omega_2 \to \infty} U_1(q) = V_{A_{N-1}}(x) + e^{x_N - x_1} = V_{A_N}^{(1)}(x).$$

Here and in what follows we shall, in calculating the limits, omit the constant C that can be made zero by subtracting the constant $N(N + 3)C$, inessential for the equations of motion, from all our potentials. The potential $U_2(q)$, when $\omega_2 \to \infty$, and with a given choice of q, τ for $N > 2$ diverges as the numbers $k < j \leq N$ may always include such for which $j + k = N + 2$, $q_j + q_k = x_j + x_k + \omega_2$, and $\wp(q_j + q_k) - C$ remains constant, while g^2 increases indefinitely. Consequently, for $N > 2$, we should put in our potential $\varepsilon = 0$, and then we arrive at a periodic Toda lattice. For $N = 2$ and $\tau = 1/2$, the expansion of $\wp(\zeta + \omega_2 \delta)$ does not hold valid and it is to be changed to the asymptotic representations

$$\wp(\zeta) \sim \frac{1}{4 \sinh^2 \zeta/2}, \quad \wp(\zeta + \omega/2) \sim e^{-\omega_2/2} \cosh \zeta,$$

$$\wp(\zeta + \omega_1/2) \sim -\frac{1}{4 \cosh^2(\zeta/2)}, \quad \wp\left(\zeta + \frac{\omega_1 + \omega_2}{2}\right) \sim -e^{-\omega_2/2} \cosh \zeta.$$

By this means, we obtain a lattice with the potential

$$V^{(2)}(x) = e^{x_1 - x_2} + e^{x_2 - x_1} + e^{x_1 + x_2} + e^{-x_1 - x_2}$$

$$+ \frac{A}{\sinh^2(x_1/2)} + \frac{B}{\sinh^2 x_1} + \frac{C}{\sinh^2(x_2/2)} + \frac{D}{\sinh x_2}.$$

When $N > 2$, there is one more possible choice in (1.113) not leading to divergences of $U_2(q)$. It is realized under the condition

$$\tau \min(j + k - 2) = 1 - \tau \max(j + k - 2), \quad N \geq j > k \geq 1.$$

From here it follows that $\tau = \frac{1}{2(N-1)}$, and at $g^2 = e^{\omega_2 \tau}$ the limits of all the terms in our potential are finite:

$$\lim_{\omega_2 \to \infty} U_1(q) = V_{A_{N-1}}(x), \quad \lim_{\omega_2 \to \infty} U_2(q) = e^{-x_1 - x_2} + e^{x_{N-1} + x_N},$$

$$\lim_{\omega_2 \to \infty} U_3(q) = \frac{\tilde{A}}{\sinh^2(x_1/2)} + \frac{\tilde{B}}{\cosh^2(x_1/2)}$$

$$+ \frac{\tilde{C}}{\sinh^2(x_N/2)} + \frac{\tilde{D}}{\cosh^2(x_N/2)},$$

which corresponds to a lattice with the potential

$$V(x) = V_{A_{N-1}}(x) + e^{-x_1 - x_2} + e^{x_{N-1} + x^N}$$

$$+ \frac{A}{\sinh^2(x_1/2)} + \frac{B}{\sinh^2 x_1} + \frac{C}{\sinh^2(x_N/2)} + \frac{D}{\sinh^2 x_N}$$

$$(1.115)$$

The constants A, B, C and D are arbitrary and connected with $\tilde{A}, \tilde{B}, \tilde{C}, \tilde{D}$ by linear relations. It is easy to see that the potential (1.115) differs from $V^{(2)}(x)$ only by the term $e^{x_2 - x_1}$ if we put $N = 2$ there. The case of $D_N^{(1)}$ in (1.114) corresponds to vanishing all four arbitrary constants in (1.115) of terms describing non-exponential interaction of the first and last particles of the lattice with external fields.

Previously we have considered the limits that allow us to derive the Hamiltonians of the Toda lattices from elliptic potentials. Integrability of new lattices requires the corresponding limits to exist for the Lax matrices, i.e. the divergences to be absent in the constants of motion of higher orders in momenta. For the lattices with potentials $V_1(x)$, $V_2(x)$, $V_3(x)$, it can be easily proved with the Lax representation found by Inozemtsev and Meshcheryakov for the trigonometric degeneration of the elliptic potentials. For the lattice (1.115) immediately obtained from elliptic potential, it is simpler to construct the Lax matrices anew, as in the representation found by Inozemtsev (see next section) who has used matrices of a higher dimension than in the paper by Inozemtsev and Meshcheryakov. Here, we shall cite the results for the lattices with potentials $V_2(x)$, $V_3(x)$ and (1.115). In all the cases, the matrices L and M obeying the Lax equation $\frac{dL}{dt} = [L, M]$ have the dimension $2N \times 2N$ and the structure

$$L = \begin{pmatrix} l & \psi \\ \psi^* & -l \end{pmatrix}, \quad M = \begin{pmatrix} m & \chi \\ -\chi^* & m \end{pmatrix}, \qquad (1.116)$$

where l, m, ψ and χ are $N \times N$ matrices, l and m being the same for almost all the lattices:

$$l_{jk} = \delta_{jk} p_j + i \left[\delta_{j,k-1} \exp\left(\frac{x_j - x_{j+1}}{2} \right) \right.$$
$$\left. + \delta_{j-1,k} \exp\left(\frac{x_k - x_{k+1}}{2} \right) \right],$$

$$m_{jk} = \delta_{jk} \mu_j + i \left[\delta_{j,k-1} \exp\left(\frac{x_j - x_{j+1}}{2} \right) \right.$$
$$\left. - \delta_{j-1,k} \exp\left(\frac{x_k - x_{k+1}}{2} \right) \right]$$

The matrices ψ, ψ^*, χ and χ^* are meromorphic functions of the spectral parameter h and for the potentials $V_2(x)$, $V_3(x)$ are of the following form (for simplicity we set $D = 1$ for $V_2(x)$ and $B = D = 1$ for $V_3(x)$):

$$\psi_{jk} = \lambda_{jk} + \sqrt{2} \left(\frac{Ch}{h^2 - 1} + h \delta_{jN} e^{x_N} \right) \delta_{jk},$$

$$\psi_{jk}^* = -\lambda_{jk} + \sqrt{2} \left(\frac{-Ch}{h^2 - 1} + h^{-1} \delta_{jN} e^{x_N} \right) \delta_{jk},$$

$$\chi_{jk} = -\frac{1}{2}\lambda_{jk} + \frac{\sqrt{2}}{2}h\delta_{jk}\delta_{jN}e^{x_N},$$

$$\chi_{jk}^* = \frac{\lambda_{jk}}{2} + \frac{\sqrt{2}}{2}\delta_{jk}\delta_{jN}\frac{e^{x_N}}{h},$$

$$\mu_j = 0, \quad \lambda_{jk} = ie^{-\frac{x_1+x_2}{2}}(\delta_{j1}\delta_{k2} + \delta_{j2}\delta_{k1}),$$

for $V_2(x)$,

$$\psi_{jk} = \sqrt{2}\delta_{jk}\left(\frac{h(Ah-C)}{h^2-1} + \delta_{j1}e^{-x_1} + h\delta_{jN}e^{x_N}\right),$$

$$\psi_{jk}^* = \sqrt{2}\delta_{jk}\left(\frac{Ch-A}{h^2-1} + \delta_{j1}e^{x_1} + h^{-1}\delta_{jN}e^{x_N}\right),$$

$$\chi_{jk} = \frac{\sqrt{2}}{2}\delta_{jk}(-\delta_{j1}e^{-x_1} + h\delta_{jN}e^{x_N}),$$

$$\chi_{jk}^* = \frac{\sqrt{2}}{2}\delta_{jk}(-\delta_{j1}e^{-x_1} + h^{-1}\delta_{jN}e^{x_N}), \quad \mu_j = 0.$$

for $V_3(x)$.

For the lattice (1.115) with four arbitrary constants, let us introduce the following notation:

$$a = \frac{\sqrt{B} + \sqrt{4A+B}}{\sqrt{2}}, \quad \lambda = \frac{-\sqrt{B} + \sqrt{4A+B}}{\sqrt{2}},$$

$$b = \frac{\sqrt{D} + \sqrt{4C+D}}{\sqrt{2}}, \quad \tau = \frac{-\sqrt{D} + \sqrt{4C+D}}{\sqrt{2}},$$

$$\rho_{jk} = (\delta_{j1}\delta_{k2} + \delta_{j2}\delta_{k1})\exp\left(-\frac{x_1+x_2}{2}\right),$$

$$\nu_{jk} = (\delta_{j,N-1}\delta_{kN} + \delta_{k,N-1}\delta_{jN})\exp\left(\frac{x_{N-1}+x_N}{2}\right).$$

The matrices $\psi, \psi^*, \chi, \chi^*$ in this notation can be represented in a relatively simple form:

$$\psi_{jk} = i\left[\delta_{jk}\left(\frac{2h(\lambda h - \tau)}{h^2-1} + \left(\lambda(\coth x_1 - 1) + \frac{a}{\sinh x_1}\right)\delta_{j1}\right.\right.$$
$$\left.\left. - \left(\tau(\coth x_N - 1) + \frac{b}{\sinh x_N}\right)h\delta_{jN}\right) + \rho_{jk} + h\nu_{jk}\right],$$

$$\psi_{jk}^* = -i \left[\delta_{jk} \left(\frac{2(\tau h - \lambda)}{h^2 - 1} + \left(\lambda(\coth x_1 - 1) + \frac{a}{\sinh x_1} \right) \delta_{j1} \right. \right.$$
$$\left. \left. - \left(\tau(\coth x_N - 1) + \frac{b}{\sinh x_N} \right) h^{-1} \delta_{jN} \right) + \rho_{jk} + h^{-1} \nu_{jk} \right],$$

$$\chi_{jk} = \frac{i}{2} \left[\delta_{jk} \left(-\frac{\lambda + a \cosh x_1}{\sinh^2 x_1} \delta_{j1} + \frac{\tau + b \cosh x_N}{\sinh^2 x_N} \delta_{jN} h \right) \right.$$
$$\left. - \rho_{jk} + h \nu_{jk} \right],$$

$$\chi_{jk}^* = -\frac{i}{2} \left[\delta_{jk} \left(-\frac{\lambda + a \cosh x_1}{\sinh^2 x_1} \delta_{j1} + \frac{\tau + b \cosh x_N}{\sinh^2 x_N} \delta_{jN} h^{-1} \right) \right.$$
$$\left. - \rho_{jk} + h^{-1} \nu_{jk} \right],$$

$$\mu_j = -i[(a + \lambda \cosh x_1)(\sinh x_1)^{-2} \delta_{j1}$$
$$+ (\tau \cosh x_N + b)(\sinh x_N)^{-2} \delta_{jN}].$$

The matrix L as a function of h has extra poles at the points $h = \pm 1$. This phenomenon is characteristic of almost all the trigonometric degenerations of the elliptic potential including those describing the motion of systems of interacting particles of the Sutherland type in an external field created by a particle of infinite mass with the potential $W(x) = A \cosh(2x) + B \cosh(x + \gamma)$ when $AB \neq 0$ which were described earlier here. For the rational degenerations of the elliptic potential, the matrix L has one pole at $h = 0$ and a pole common with M at infinity. On each of $2N$ sheets of the spectral curve $\det(L(h) - I_z) = 0$, the eigenvectors of L possess essential singularities at the poles of matrix M, which allows one to construct the vector Baker–Akhiezer functions and to obtain explicit expressions for particle trajectories as combinations of multidimensional theta functions. To the best of our knowledge, this program is performed only for the situation described in the previous section.

The above consideration may also be applied to "relativistic" analogs of the rational, trigonometric and elliptic potentials whose integrability at $\varepsilon = 0$ has been established by Ruijsenaars and at $\varepsilon = 1$, $N = 2$ by Inozemtsev and Sasaki.

Note also that the lattices (1.115) possess non-Abelian generalizations: the systems of nonlinear matrix equations for matrices $g_j \in gl(s, \mathbf{R})$, $1 \le j \le N$,

$$\frac{d}{dt}\left(\frac{dg_j}{dt}g_j^{-1}\right) = g_{j-1}g_j^{-1} - g_jg_{j+1}^{-1} + \delta_{j1}(g_1^{-2} + \alpha g_1^{-1}) - \delta_{jN}(g_N^2 + \beta g_N)$$

admit· the Lax representation with the structure of L and M (1.116) where l, m, ψ and χ are block $(N_s \times N_s)$ matrices:

$$l_{jk} = \delta_{jk}\frac{dg_j}{dt}g_j^{-1} + \delta_{j,k-1}g_jg_k^{-1} + \delta_{j-1,k},$$

$$m_{jk} = \frac{1}{2}\left(-\delta_{jk}\frac{dg_j}{dt}g_j^{-1} + \delta_{j,k-1}g_jg_k^{-1} - \delta_{j-1,k}\right),$$

$$\psi_{jk} = \frac{\delta_{jk}}{\sqrt{2}}\left(\frac{2(\alpha h - \beta)h}{h^2 - 1} + g_1^{-1}\delta_{j1} + h_{g_N}\delta_{jN}\right),$$

$$\psi_{jk}^* = \frac{\delta_{jk}}{\sqrt{2}}\left(\frac{2(\beta h - \alpha)}{h^2 - 1} + g_1^{-1}\delta_{j1} + h^{-1}g_N\delta_{jN}\right),$$

$$\chi_{jk} = \frac{1}{2\sqrt{2}}\delta_{jk}[-g_1^{-1}\delta_{j1} + h_{g_N}\delta_{jN}],$$

$$\chi_{jk}^* = \frac{1}{2\sqrt{2}}\delta_{jk}[-g_1^{-1}\delta_{j1} + g_N h^{-1}\delta_{jN}].$$

These systems represent generalizations of the non-Abelian lattices found by Reyman and analogous of Abelian systems of the type $C_N^{(1)}$ and $D_{N+1}^{(2)}$. Two-dimensional existing for all earlier known lattices are absent for the systems with potentials $V_2(x)$ and $V_3(x)$ which cannot be included into the Zakharov–Shabat scheme of construction of two-dimensional nonlinear evolution equations. It is to be noted that of extreme interest seems to be the group-theoretical interpretation of the systems with potentials $V_2(x)$, $V_3(x)$. For the previously known lattices, it has been obtained by Kostant and Symes, where a detailed study has been made for the connection of the Lax matrices with orbits of a coadjoint representation of the Kac–Moody algebras, and the solution of equations of motion has been reduced to the problem of factorization in the corresponding infinite-dimensional groups. For the potentials $V_1(x)$, $V_2(x)$ with the exponential interaction, the Lax matrices on the gauge $\tilde{L} = \Omega L \Omega^-$, $\tilde{M} = \Omega M \Omega^{-1}$,

$$\Omega = \begin{pmatrix} E & 0 \\ 0 & F \end{pmatrix}, \quad E_{jk} = \delta_{jk}, \quad F_{jk} = \delta_{j,N-k+1}$$

can also be realized as orbits of the Kac–Moody algebras. As for the lattices (1.115), it is necessary first to obtain the algebraic interpretation of the potential containing Weierstrass functions. At present there is no such interpretation even for the simplest case of $\varepsilon = 0$.

1.10. Lax Representation with Spectral Parameter on a Torus for Integrable Particle Systems

In this section, we shall study the Hamiltonian (1.103) with elliptic potentials:

$$V(x) = g^2 \wp(x),$$
$$W(x) = g_1^2 \wp(x) + g_2^2 \wp\left(x + \frac{\omega_1}{2}\right) + g_3^2 \wp\left(x + \frac{\omega_2}{2}\right)$$
$$+ g_4^2 \wp\left(x + \frac{\omega_1 + \omega_2}{2}\right).$$

with the Weierstrass function $\wp(x)$ having periods ω_1, ω_2. In Section 1.8, we found generalization of the results of Olshanetsky and Perelomov, but the constants g_α obey one or two nonlinear equations defining either three- or two-dimensional hypersurfaces in a four-dimensional space $\{g_\alpha\}$. The Lax matrices did not depend on a spectral parameter, which can be easily introduced for generalized Toda lattices, which are limits of (1.103). Now, we determine the L, M pair for the Hamiltonian (1.103) with elliptic potentials that allow one to remove the above-mentioned constraints on the constants $\{g_\alpha\}$ and, what is more, depend on the spectral parameter. This Hamiltonian, which depends on seven arbitrary constants, is the most general one of the family with the structure (1.103).

To start with, we take the matrices L and M in the form resembling that used in Section 1.8,

$$L = \begin{pmatrix} l & \lambda & \psi \\ \lambda & 0 & -\lambda \\ -\psi & -\lambda & -l \end{pmatrix},$$

$$M = \begin{pmatrix} m & \omega & s \\ -\omega & \mu & -\omega \\ s & \omega & m \end{pmatrix} \tag{1.116a}$$

$(l, \psi, \lambda, m, \omega$ and μ are now matrices of $N \times N$ dimension). The rank of L equals $2N$ as its eigenvectors corresponding to the zeroth eigenvalue compose a linear N-dimensional space. The first and last N coordinates of these vectors coincide and are arbitrary, $y_j = y_{2N+j}$, the other coordinates are calculated by the formula

$$y_{j+N} = -\{\lambda^{-1}(l + \psi)\}_{jk} y_k, \quad 1 \le j, k \le N.$$

The matrices l, ψ, s and m are almost of the same structure as in Section 1.8:

$$l_{jk} = p_j \delta_{jk} + i(1 - \delta_{jk}) g R(x_j - x_k),$$

$$\psi_{jk} = i[\delta_{jk}\nu(x_j) + (1 - \delta_{jk}) g R(x_j + x_k)],$$

$$m_{jk} = i\left[\delta_{jk}\left(\tau(x_j) - \sum_{n \ne j}(z(x_j - x_n) + z(x_j + x_n)) \right) \right.$$

$$\left. + (1 - \delta_{jk}) g R'(x_j - x_k) \right],$$

$$s_{jk} = i\left[\frac{\delta_{jk}}{2}\nu'(x_j) + (1 - \delta_{jk}) g R'(x_j + x_k) \right].$$

The principal difference from the Ansatz used in Section 1.8 consists of the structure of the matrices λ, ω and μ that here are of $N \times N$ dimension. The first two of them are diagonal:

$$\lambda_{jk} = \lambda(x_j)\delta_{jk}, \quad \omega_{jk} = \lambda'(x_j)\delta_{jk}.$$

Hereafter, the prime means differentiation of a function with respect to its argument. The Lax equation $\frac{dL}{dt} = [L, M]$, upon substitution into it of (1.116a), splits into four equations for the derivatives of l, λ and ψ

$$\frac{dl}{dt} = [l, m] + \{s, \psi\} - \{\lambda, \omega\}, \tag{1.117}$$

$$\frac{d\psi}{dt} = [\psi, m] + \{l, s\},$$

$$\frac{d\lambda}{dt} = (l + \psi)\omega + \lambda\mu - (m - s)\lambda$$

$$= \lambda(m - s) - \mu\lambda + \omega(l - \psi). \tag{1.118}$$

As the Hamiltonian (1.103) is connected with L by $H = \frac{1}{4}\operatorname{Tr} L^2$, we obtain for the functions $V(x)$ and $W(x)$ the equations

$$V(x) = R^2(x), \quad W(x) = \frac{\nu^2}{2}(x) + \lambda^2(x).$$

Equations (1.117) have the same solutions as it were found in Section 1.8,

$$R(x) = \frac{a}{\operatorname{sn} ax}, \quad z(x) = -\frac{ga^2}{\operatorname{sn}^2 ax}, \quad \tau(x) = \frac{a\nu(x)}{\operatorname{sn} 2\, ax},$$

$$\nu(x) = a\frac{\alpha + \beta \operatorname{sn}^2 ax + \gamma \operatorname{sn}^4 ax}{\operatorname{sn} ax \operatorname{cn} ax \operatorname{dn} ax},$$

the constants a, α, β, γ and modulus of the Jacobi elliptic functions are arbitrary.

Now, consider the Equation (1.118). As diagonal elements of the matrices in (1.118) are equal, we can express μ_{jj} in terms of q, z, ν, τ and λ:

$$\mu_{jj} = i\left[\tau(x_j) - \sum_{n\neq j}(z(x_j - x_n) + z(x_j + x_n)) - \frac{d\nu(x_j)}{2dx_j}\right].$$

Non-diagonal elements of (1.118) give two equations for μ_{jk}:

$$\mu_{jk} = [(m_{jk} - s_{jk})\lambda(x_k) - (l_{jk} + \psi_{jk})\lambda'(x_k)][\lambda(x_j)]^{-1}$$
$$= [(m_{jk} - s_{jk})\lambda(x_j) - (l_{jk} - \psi_{jk})\lambda'(x_j)][\lambda(x_k)]^{-1}.$$

The conditions of their compatibility is a functional equation for λ. Denoting $\lambda^2(x) = \chi(x)$, we write it as follows:

$$R(x - y)[\chi'(x) + \chi'(y)] + R(x + y)[\chi'(y) - \chi'(x)]$$
$$+ 2[R'(x - y) - R'(x + y)][\chi(x) - \chi(y)] = 0$$

or, upon substituting of $R(x \pm y)$ in elliptic form,

$$\chi'(x)[\operatorname{sn} a(x + y) - \operatorname{sn} a(x - y)] + \chi'(y)[\operatorname{sn} a(x + y) + \operatorname{sn} a(x - y)]$$
$$= 2a[\chi(x) - \chi(y)][\operatorname{sn} a(x + y) \operatorname{sn} a(x - y)]^{-1}$$
$$\times [\operatorname{cn} a(x - y) \operatorname{dn} a(x - y) \operatorname{sn}^2 a(x + y)$$
$$- \operatorname{cn} a(x + y) \operatorname{dn} a(x + y) \operatorname{sn}^2 a(x - y)].$$

Using the addition formula for the Jacobi elliptic functions and making lengthy but not complicated calculations, we reduce this formula to the form

$$\chi'(x)\frac{\operatorname{cn} ax\operatorname{dn} ax}{\operatorname{sn} ax} + \chi'(y)\frac{\operatorname{cn} ay \operatorname{dn} ay}{\operatorname{sn} ay}$$
$$= \frac{2a(\chi(x) - \chi(y))}{\operatorname{sn}^2 ax - \operatorname{sn}^2 ay}(\operatorname{cn}^2 ax \operatorname{dn}^2 ay + \operatorname{cn}^2 ay \operatorname{dn}^2 ax).$$

Now, let us look for a general solution to this equation. As the right-hand side of that equation cannot have a pole for all x when $y \to 0$, we have

$$\chi(y) \sim \delta_0 + \frac{\delta a^2}{2}y^2,$$

for small y, and in the limit $y \to 0$, we obtain

$$\chi'(x)\frac{\operatorname{cn} ax \operatorname{dn} ax}{\operatorname{sn} ax} + a\delta = \frac{2a(\chi(x) - \delta_0)}{\operatorname{sn}^2 ax}(\operatorname{cn}^2 ax + \operatorname{dn}^2 ax),$$

i.e. a linear equation of the first order with the solution

$$\chi(x) = \delta_0 + \frac{\delta}{2}\frac{\mathrm{sn}^2 ax}{\mathrm{cn}^2 ax\, \mathrm{dn}^2 ax} + \delta_1 \frac{\mathrm{sn}^4 ax}{\mathrm{cn}^2 ax\, \mathrm{dn}^2 ax},$$

where δ_1 is an integration constant. Direct substitution of the solution into our functional equation shows that no extra constraints arise on the constants δ_0, δ and δ_1 and we found just the required general solution to the functional equation. Since δ_0 gives only a negligible addition to the potential $W(x)$, that constant may be chosen so that $\lambda(x)$ be of the form

$$a^2[(\alpha_1^2 \,\mathrm{cn}^4 ax + \alpha_2^2 \,\mathrm{dn}^4 ax)(\mathrm{cn}\, ax\, \mathrm{dn}\, ax)^{-2} + 2\alpha_1\alpha_2],$$

or

$$\lambda(x) = a\frac{\alpha_1 \,\mathrm{cn}^2 ax + \alpha_2 \,\mathrm{dn}^2 ax}{\mathrm{cn}\, ax\, \mathrm{dn}\, ax}.$$

Going back to the formula for $W(x)$, we see that now there is a sufficient number of parameters to ensure the absence of any constraints on the constants g_1, g_2, g_3 and g_4 in the formula for $W(x)$. And, what is more, their number is larger by one than the number of $\{g_\alpha\}$, therefore a certain combination of $\alpha, \beta, \gamma, \alpha_1, \alpha_2$ can be used as a spectral parameter.

Let us transform the formula for $W(x)$ with the functions ν and λ to the form of Weierstrass functions, i.e. express the squares of the Jacobi elliptic functions through the Weierstrass ones. This can be achieved with the relations

$$\wp(x) = \frac{a^2}{\mathrm{sn}^2 ax} + c_0, \qquad \wp\left(x + \frac{\omega_1}{2}\right) = \frac{a^2 \,\mathrm{dn}^2 ax}{\mathrm{cn}^2 ax} + c_0,$$

$$\wp\left(x + \frac{\omega_2}{2}\right) = -a^2 \,\mathrm{dn}^2 ax + c_0 + a^2, \qquad (1.119)$$

$$\wp\left(x + \frac{\omega_1 + \omega_2}{2}\right) = k^2 a^2 \frac{\mathrm{cn}^2 ax}{\mathrm{dn}^2 ax} + c_0,$$

where k is the modulus of the Jacobi elliptic functions,

$$c_0 = -\frac{a^2(1 + k^2)}{3},$$

and the periods ω_1 and ω_2 of the function $\wp(x)$ will be used as independent parameters instead of a and k.

Taking advantage of $\alpha, \beta, \gamma, \alpha_1$ and α_2 being arbitrary, we write the functions $\lambda(x)$ and $\nu(x)$ in the form

$$\lambda(x) = a\frac{\tilde{\alpha}_1 k^2 \operatorname{cn}^2 ax + \tilde{\alpha}_2 \operatorname{dn}^2 ax}{\operatorname{cn} ax \operatorname{dn} ax},$$

$$\nu(x) = a\sqrt{2}\frac{\tilde{\alpha} \operatorname{cn}^2 ax \operatorname{dn}^2 ax + \tilde{\beta} \operatorname{dn}^2 ax - \tilde{\gamma}k^2 \operatorname{cn}^2 ax \operatorname{sn}^2 ax}{\operatorname{sn} ax \operatorname{cn} ax \operatorname{dn} ax}$$

(1.120)

which allows us to find a simple connection between the constants g_1, g_2, g_3, g_4 and the Lax matrix parameters. Substituting (1.119) into the formula for $W(x)$ and using (1.120), we obtain from the comparison of two formulas for $W(x)$ the following system of equations quadratic in $\tilde{\alpha}, \tilde{\beta}, \tilde{\gamma}, \tilde{\alpha}_1$ and $\tilde{\alpha}_2$:

$$(\tilde{\alpha} + \tilde{\beta})^2 = g_1^2, \quad (\tilde{\alpha} + \tilde{\gamma})^2 = g_3^2,$$

$$\tilde{\beta}^2 + \tilde{\alpha}_2^2 = g_2^2, \quad \tilde{\gamma}^2 + \tilde{\alpha}_1^2 = g_4^2.$$

From these equations, we may express $\tilde{\alpha}, \tilde{\beta}$ and $\tilde{\gamma}$ through $\{g_\alpha\}, \tilde{\alpha}_1$ and $\tilde{\alpha}_2$:

$$\tilde{\alpha} = \frac{1}{2}(g_1 + g_3) - \frac{1}{2}(g_1 - g_3)^{-1}[\tilde{\alpha}_1^2 - \tilde{\alpha}_2^2 - g_2^2 - g_4^2],$$

$$\tilde{\beta} = \frac{1}{2}(g_1 - g_3) + \frac{1}{2}(g_1 - g_3)^{-1}[\tilde{\alpha}_1^2 - \tilde{\alpha}_2^2 + g_2^2 - g_4^2],$$

$$\tilde{\gamma} = -\frac{1}{2}(g_1 - g_3) + \frac{1}{2}(g_1 - g_3)^{-1}[\tilde{\alpha}_1^2 - \tilde{\alpha}_2^2 + g_2^2 - g_4^2]$$

whereas $\tilde{\alpha}_1$ and $\tilde{\alpha}_2$ are related by

$$(\tilde{\alpha}_1^2 - \tilde{\alpha}_2^2)^2 + 2A\tilde{\alpha}_1^2 + 2B\tilde{\alpha}_2^2 + C = 0,$$

$$A = g_2^2 - g_4^2 + (g_1 - g_3)^2,$$

$$B = -g_2^2 + g_4^2 + (g_1 - g_3)^2,$$

$$C = (g_2^2 - g_4^2)^2 - 2(g_2^2 + g_4^2)(g_1 - g_3)^2 + (g_1 - g_3)^4.$$

(1.121)

Equation (1.120) defines an algebraic curve of the fourth order. It is easily verified that its genus equals one and, consequently, its elliptic uniformization is feasible. To determine the explicit form of the functions accomplishing this uniformization, we substitute into (1.120) the following expressions:

$$\tilde{\alpha}_1 = \frac{\delta\,\text{sn}\,h\,\text{cn}\,\varphi\,\text{dn}\,\varphi}{1 - \kappa^2\,\text{sn}^2\,h\,\text{sn}^2\,\varphi},$$

$$\tilde{\alpha}_2 = \frac{\delta\,\text{cn}\,h\,\text{dn}\,h\,\text{sn}\,\varphi}{1 - \kappa^2\,\text{sn}^2\,h\,\text{sn}^2\,\varphi}, \tag{1.122}$$

where κ is the modulus of elliptic functions in these formulas. Upon rather lengthy computations, it can be shown that (2.121) is fulfilled for arbitrary values of h if δ, κ and φ are connected with A, B and C as follows:

$$\text{sn}^2\varphi = -(A+B)^{-1}[-B + \sqrt{B^2 - C}]$$
$$\times\,[1 + (AB - \sqrt{(C - A^2)(C - B^2)})C^{-1}],$$

$$\delta = (-B + \sqrt{B^2 - C})^{1/2}(\text{sn}\,\varphi)^{-1},$$

$$\kappa = C^{1/2}\,\text{sn}^2\varphi(-B + \sqrt{B^2 - C}). \tag{1.123}$$

Equations (1.122) and (1.123) give explicit uniformization of the curve (1.120) and h is a spectral parameter. The dependence of the Lax matrices on h is determined by the formulas for $\lambda(x)$, $\nu(x)$ and (1.120). The range of variation of h is a complex torus that is a factor of the plane $\mathbf{C}$ with respect to the lattice of periods Γ of the function $\text{sn}\,h$ with the modulus (1.120). Note that for trigonometric degeneration of the elliptic potentials, the Lax matrices are of $(2N \times 2N)$ dimension, and L, M are meromorphic functions of the spectral parameter with poles at four points of the plane $\mathbf{C}$, $h = 0, \pm 1, \infty$. In the considered elliptic case, the poles of L and M as functions of h are at the points $\text{sn}\,h = \pm(\kappa\,\text{sn}\,\varphi)^{-1}$ and at the points of poles of $\text{sn}\,h$ belonging to $\mathbf{C}/\Gamma$.

The spectral equation $\det(L(h) - wE) = 0$ (E is a unit $3N \times 3N$ matrix) defines the curve $w(h)$ covering $\mathbf{C}/\Gamma$. The number of sheets of the covering coincides with the rank of L and equals $2N$. It is certainly of interest to consider the possibility of linearization of the Hamiltonian flow (1.103) with elliptic potentials on the Jacobi variety of the curve $w(h)$. Investigation of the rational degenerations in Section 1.6 shows that this sort of linearization is feasible. Specifically, integration of the equations of motion of systems with the Hamiltonian (1.103) and its further degenerations can be made by the methods of algebraic geometry developed by Dubrovin and Krichever. However, up to our best knowledge, nobody has been able to perform this program to date.

Chapter 2

Quantum Systems

2.1. The Ground-State Wave Functions of Calogero–Sutherland Quantum Systems in an External Field

The notion of quantum integrability is as well defined as the classical one due to the absence of quantum Liouville theorem. It is generally believed that quantum integrability takes place if there is a commutative ring $\{I\}$ with number of quantum operators equal to the number of degrees of freedom of the system under consideration, which includes Hamiltonian operator. For the systems studied in Chapter 1, these rings are still not found with the exception of simplest system with $N = 2$ where the Hamiltonian is

$$H = \frac{p_1^2 + p_2^2}{2} + \frac{g^2}{(x_1 - x_2)^2} + f^2(x_1^4 + x_2^4)$$

(here and in all of Chapter 2 the symbol $\{p_j\}$ means differential operator $\{-\frac{\partial}{\partial x_j}\}$). The second member of the commutative ring looks like

$$I = p_1^2 p_2^2 + f^2(p_1^2 x_2^4 + p_2^2 x_1^4) - p_1 \frac{g^2}{(x_1 - x_2)^2} p_2$$

$$- p_2 \frac{g^2}{(x_1 - x_2)^2} p_1 + \frac{2g^2}{(x_1 - x_2)^2} f^2 x_1^2 x_2^2$$

$$+ \frac{g^4}{(x_1 - x_2)^4} + f^4 x_1^4 x_2^4.$$

Japanese mathematicians H. Ochiai *et al.* [29] found a family of differential operators commuting with the most general Hamiltonian resembling the classical elliptic Hamiltonian described in Section 1.10 of Chapter 1 but were not able to prove that they form a commutative ring.

Nevertheless, it is generally believed that there is a deep analogy between the integrable classical systems of particles and quantum ones. In particular, it turns out that there it will be possible to find at least part of the eigenfunctions of the Schrödinger operator. At the outset, let us start with the quantum equation

$$\left(\sum_{j=1}^{N} \left[-\frac{1}{2}\partial^2/\partial x_j^2 + W(x_j) \right] + \sum_{j>k} V(x_j - x_k) \right) \psi = E\psi, \quad (2.1)$$

where $\psi(x_1, \ldots, x_N)$ is the wave function of the system under consideration. When $W(x) \neq 0$, the ground-state function may also be factorizable as Sutherland found for the case $W(x) = 0$:

$$\psi_0(x_1, \ldots, x_N) = \prod_{j=1}^{N} c(x_j) \prod_{j>k} \chi(x_j - x_k). \quad (2.1a)$$

Let us find the conditions for solutions of Equation (2.1) to possess property (2.1a). Substitution of (2.1a) into (2.1) leads also to the Sutherland equation for $\varphi(x)$, the logarithmic derivative of the function $\chi(x)$:

$$\varphi(x)\varphi(y) + [\varphi(x) + \varphi(y)]\varphi(-x - y) = f(x) + f(y) + f(x + y),$$

where $f(x)$ is some (still undetermined) function. The solutions which have been found by Sutherland, are

$$\varphi(x) = ax + b/x, \quad f(x) = -ab - 1/2a^2x^2,$$

$$\varphi(x) = a\coth(bx), \quad f(x) = -\frac{1}{3}a^2.$$

Its general solution was found by Calogero as follows:

$$\varphi(x) = \alpha\zeta(x) + \beta x,$$

$$f(x) = \frac{1}{2}[\alpha\beta - \alpha\varphi'(x) - \varphi^2(x)],$$

where $\zeta(x)$ is the Weierstrass ζ function.

To satisfy (2.1) for $W(x) \neq 0$, this condition is not sufficient. Denoting $\tau(x) = c'(x)/c(x)$ the logarithmic derivative of the function $c(x)$, we arrive at the functional equation connecting $\varphi(x)$ and $\tau(x)$:

$$\varphi(x - y)[\tau(x) - \tau(y)] = \lambda(x) + \lambda(y).$$

The potential $W(x)$ is expressed in terms of $\tau(x)$ and $\lambda(x)$ up to a constant, as follows:

$$W(x) = \frac{1}{2}[\tau'(x) + \tau^2(x)] + (N - 1)\lambda(x) + \text{const.}$$

The last functional equation possesses a much smaller class of solutions than equation for φ. All its analytic solutions can be obtained by a standard method, by expanding it in powers of $(x - y)$ and taking account of possible singularities of $\varphi(x)$ at the point $x = 0$. In particular, the function $\varphi(x)$ is defined up to two arbitrary constants:

$$\varphi(x) = \alpha a \coth(ax).$$

It is a partial solution of the general Sutherland equation. The other functional equation gives

$$\tau(x) = -b\sinh(2ax) + b_1, \quad \lambda(x) = \alpha a b \cosh(2ax),$$

where b and b_1 are constants, and in order that the wave function decreases when $|x_j - x_k| \to \infty$, we should take $b > 0$. Thus, all the potentials $V(x)$ and $W(x)$ for which there exist solutions to the Schrödinger equation of type (2.1a) are completely determined:

$$V(x) = \alpha(\alpha - 1)a^2/\sinh^2(ax),$$

$$W(x) = \frac{1}{4}b^2\cosh(4ax) - bb_1\sinh(2ax)$$

$$- ab[1 + \alpha(N - 1)]\cosh(2ax). \tag{2.2}$$

Note, first, that the Hamiltonian of an N-particle system with potentials (2.2) is hermitean only when $\alpha(\alpha - 1) > 3/4$. The normalization integral for a wave function ψ_0 (2.1a)

$$\psi_0(x_1, \ldots, x_N) = C_N \prod_{j=1}^{N} \exp[-(b/2a)\cosh(2ax_j) + b_1 x_j]$$

$$\times \prod_{j>k}^{N} |\sinh a(x_j - x_k)|^{\alpha} \tag{2.3}$$

converges only for positive values of α defined by the above inequality. The function $\psi_0(x_1, \ldots, x_N)$ has no zeros outside the hyperplanes $x_j - x_k = 0$ which contain singularities of the Hamiltonian, therefore, following Calogero, we may conclude that the function describes the ground state of a light particle system in a field created by a particle of infinite mass. The energy of this state is calculated by substituting (2.3) and (2.1a) into Equation (2.1) as follows:

$$E_0 = \frac{1}{2} N \left[\frac{1}{3} \alpha^2 a^2 (N^2 - 1) + b_1^2 - \frac{1}{2} b^2 \right].$$

The spectrum of excited states for $b \neq 0$ is discrete.

Secondly, classical systems of particles the interaction of which is defined by the potentials $W(x)$ and $V(x)$ (2.2) are completely integrable (see Chapter 1). Extra classical integrals of motion are coefficients of the polynomial $P(\lambda) = \det|L(p_j, x_j) - \lambda E|$, where the Lax matrix L found by Inozemtsev depends on the coordinates and momenta of the particles, and E is a unit matrix. In the quantum case also for any N there may exist N integrals of motion which can be obtained from the classical ones by some ordering of the operators p_j and functions of the coordinates. In the simplest non-trivial case $N = 2$, when the Hamiltonian has the form

$$H = \frac{1}{2}(p_1^2 + p_2^2) + g^2 a^2 / \sinh^2 a(x_1 - x_2)$$

$$+ \sum_{j=1}^{2} \cosh(2ax_j + \beta)[\alpha \cosh(2ax_j + \gamma)],$$

the above problem of ordering is trivially solved, and for extra quantum integral of motion one may obtain a sufficiently simple expression

$$I = p_1^2 p_2^2 - p_1 g^2 a^2 / \sinh^2 a(x_1 - x_2) p_2$$
$$- p_2 g^2 a^2 / \sinh^2 a(x_1 - x_2) p_1 + g^4 a^4 / \sinh^4 a(x_1 - x_2)$$
$$+ p_1^2 \cosh(2_{ax_2} + \beta)[\alpha \cosh(2_{ax_2}) + \gamma]$$
$$+ p_2^2 \cosh(2ax_1 + \beta)[\alpha \cosh(2ax_1) + \gamma]$$
$$+ g^2 a^2 / \sinh^2 a(x_1 - x_2)[\cosh(2ax_2 + \beta)[\alpha \cosh(2ax_1) + \gamma]$$
$$+ \cosh(2ax_1 + \beta)[\alpha \cosh(2ax_2) + \gamma]]$$
$$+ \cosh(2ax_1 + \beta)\cosh(2ax_2 + \beta)[\alpha \cosh(2ax_1) + \gamma]$$
$$\times [\alpha \cosh(2ax_2) + \gamma].$$

An interesting particular case of (2.2) represents the Morse potential:

$$W(x) = 2a^2 A^2 (e^{4ax} - 2e^{2ax}).$$

This potential is derived from (2.2) by shifting coordinates by a number ϵ and taking the limit

$$\epsilon \to \infty, b^2 e^{4\epsilon a} \to 16a^2 A^2, b_1 \to a[2A - 1 - \alpha(N - 1)].$$

In this case, Equation (2.1) describes also the states of scattering (for instance, scattering of M particles on a bound state of $(N - M)$ particles). The Hamiltonian spectrum contains only a finite number of discrete values. The wave function (2.1a) acquires the form

$$\psi_0(x1, \ldots, x_N) = C_N \prod_{j=1}^{N} \exp(-Ae^{2ax_j} + ax_j[2A - 1 - \alpha(N - 1)])$$

$$\times \prod_{j>k}^{N} |\sinh a(x_j - x_k)|^{\alpha}. \tag{2.4}$$

It can be shown that the integral $\int |\psi_0|^2 dx_1, \ldots, dx_N$ converges provided that

$$\alpha(N - 1) < A - 1/2.$$

The wave function (2.4) is a wave function of the ground state of a system with energy

$$E_0 = -\frac{1}{2}Na^2\left[\frac{1}{3}\alpha^2(N^2-1) + (2A-1-\alpha(N-1))^2\right].$$

The physical meaning of the above inequality is clear: only a finite number of repelling particles can be "placed" in a potential well of finite depth. For $A < 1/2$, the discrete spectrum disappears; for $\alpha(N-1) > A - 1/2$, it may happen that there exist N-particle bound states the wave functions of which are no longer factorizable. At present we cannot state that such states do not exist. In the limit $\alpha \to 0$ the function (2.3) turns into a product of ground-state wave functions of non-interacting particles in the Morse potential.

For a two-particle system, we have succeeded in finding explicitly the normalization constant in (2.3):

$$C_2 = a\left(\frac{1}{2}\sqrt{\pi}(4A)^{2(1+\alpha-2A)}\frac{\Gamma(4A-2\alpha-2)\Gamma(2A-2\alpha-1)\dfrac{\Gamma(2\alpha+1)}{\Gamma(2A-\alpha-1/2)\Gamma(\alpha+1)}}{}\right)^{-1/2}$$

Factorization of some solutions to the Schrödinger equation may also hold for more general Hamiltonians describing the motion of $2N$ light particles in the field of the particle of infinite mass. This case will be considered in the next section.

2.2. Factorization of the Ground-State Wave Functions of the System of Particles Interacting with Particle of Infinite Mass

Above we have shown that the ground-state wave function is factorized in the case

$$V(x) = g^2\sinh^{-2}(ax), \quad W(x) = A\cosh(4ax)$$
$$+ B\sinh(2ax) + C\cosh(2ax).$$

Here, we shall consider the Hamiltonians of a more general form

$$H = \sum_{j=1}^{N} \left[\frac{1}{2}p_j^2 + W(x_j) \right] + \sum_{j>k} [V(x_j - x_k) + V(x_j + x_k)],$$

(2.5)

appearing in analogy with classical integrable systems (see Chapter 1). Consider the conditions under which solutions to the Schrödinger equation

$$\left(\sum_{j=1}^{N} \left[-\frac{1}{2}\partial^2/\partial x_j^2 + W(x_j) \right] + \sum_{j>k} [V(x_j - x_k) + V(x_j + x_k)] \right)$$
$$\times \psi_0 = E_0\psi_0,$$

(2.6)

are factorizable

$$\psi_0(x_1, \ldots, x_N) = \prod_{j=1}^{N} q(x_j) \prod_{j>k}^{N} c(x_j - x_k)c(x_j + x_k).$$

Let us introduce the notation

$$[\log c(\xi)]' = \alpha(\xi), \quad [\log q(\xi)]' = \gamma(\xi).$$

We assume that the functions $c(\xi)$ and $q(\xi)$ are even, whereas the functions $\alpha(\xi)$ and $\gamma(\xi)$ are odd. The substitution of ansatz for ψ_0 into (2.5) results in the following functional equation with four unknown functions:

$$[\gamma(\xi) - \gamma(\eta)]\alpha(\xi - \eta) + [\gamma(\xi) + \gamma(\eta)]\alpha(\xi + \eta)$$
$$= \tau(\xi) + \tau(\eta) + \beta(\xi - \eta) + \beta(\xi + \eta).$$

(2.7)

Differentiating both sides of this equation successively with respect to ξ and η and applying the operation $\partial^2/\partial\xi^2 - \partial^2/\partial\eta^2$, we get the equation for two functions

$$[\nu(\xi+\eta) - \nu(\xi-\eta)][\rho''(\xi) - \rho''(\eta)] + 2[\nu''(\xi+\eta)$$
$$- \nu''(\xi-\eta)][\rho(\xi) - \rho(\eta)]$$
$$+ 3\nu'(\xi+\eta)[\rho'(\xi) - \rho'(\eta)]$$
$$- 3\nu'(\xi-\eta)[\rho'(\xi) + \rho'(\eta)] = 0,$$

where

$$\nu(\xi) = \alpha'(\xi), \quad \rho(\xi) = \gamma'(\xi).$$

This equation coincides in the form with a functional equation (1.12) occuring in the study of two-particle classical systems. Its solutions are of the following form:

(i)

$$\nu(\xi) = \tilde{\lambda}_1 a^2 \sinh^{-2}(a\xi) + \tilde{\lambda}_2 a^2 \sinh^{-2}(a\xi/2),$$
$$\rho(\xi) = \tilde{\lambda}_3 a^2 \cosh(2a\xi),$$

(ii)

$$\nu(\xi) = \tilde{\lambda}_1 \wp(\xi) + \tilde{\lambda}_2,$$
$$\rho(\xi) = \tilde{\lambda}_3 \wp(\xi) + \tilde{\lambda}_4 \wp(\xi + \omega_1) + \tilde{\lambda}_5 \wp(\xi + \omega_2)$$
$$+ \tilde{\lambda}_6 \wp(\xi + \omega_1 + \omega_2) + \mu_1,$$

where $\tilde{\lambda}_1, \ldots, \tilde{\lambda}_6, \mu_1$ are arbitrary constants, and ω_1 and ω_2 are half-periods of the Weierstrass function $\wp(\xi)$.

Substituting this into functional equation (2.7), we obtain the following solutions to this equation:

(i)

$$\alpha(\xi) = \tilde{\lambda}_1 a\coth(a\xi) + \tilde{\lambda}_2 a\coth(a\xi/2), \quad \gamma(\xi) = \tilde{\lambda}_3 a\sinh(2a\xi),$$
$$\tau(\xi) = 2(\tilde{\lambda}_1 + \tilde{\lambda}_2)\tilde{\lambda}_3 a^2 \cosh(2a\xi), \quad \beta(\xi) = 2\tilde{\lambda}_2 \tilde{\lambda}_3 a^2 \cosh(a\xi),$$

(ii)

$$\alpha(\xi) = \lambda\zeta(\xi) + \mu\xi, \quad \gamma(\xi) = \tilde{\lambda}\zeta(\xi) + \tilde{\lambda}_1\zeta_1(\xi)$$
$$+ \tilde{\lambda}_2\zeta_2(\xi) + \tilde{\lambda}_3\zeta_3(\xi) + \mu_1\xi,$$
$$\tau(\xi) = 2\mu\xi\zeta(\xi) - \mu\mu_1\xi^2 + \lambda[\tilde{\lambda}\zeta^2(\xi) - \wp(\xi)]$$
$$+ \tilde{\lambda}_1[\zeta_1^2(\xi) - \wp_1(\xi)]$$
$$+ \tilde{\lambda}_2[\zeta_2^2 - \wp_2(\xi)] + \tilde{\lambda}_3[\zeta_3^2(\xi) - \wp_3(\xi)]],$$
$$\beta(\xi) = \frac{1}{2}\mu\mu_1\xi^2 + \lambda\mu_1\xi\zeta(\xi) + \frac{1}{2}\lambda(\tilde{\lambda} + \tilde{\lambda}_1 + \tilde{\lambda}_2 + \tilde{\lambda}_3)[\zeta^2(\xi) - \wp(\xi)].$$

Here, $\zeta(\xi)$ is the Weierstrass ζ function, and

$$\zeta_i(\xi) = \zeta(\xi + \omega_i) - \zeta(\omega_i), \quad \wp_i(\xi) = \wp(\xi + \omega_i),$$
$$i = 1, 2, 3, \quad \omega_3 = \omega_1 + \omega_2,$$

while $\lambda, \mu, \mu_1, \tilde{\lambda}, \tilde{\lambda}_1, \tilde{\lambda}_2, \tilde{\lambda}_3$ are arbitrary constants.

It may be verified that the solution (i) determines the ground-state wave function only when $\tilde{\lambda}_1 = 0$ or $\tilde{\lambda}_2 = 0$, i.e. when it is reduced to the solution (ii).

Now consider this solution. The potentials $V(\xi)$ and $W(\xi)$ are defined by the functions $\alpha(\xi), \beta(\xi), \gamma(\xi)$ and $\tau(\xi)$ up to arbitrary constants:

$$V(\xi) = \alpha'(\xi) + (N-1)\alpha^2(\xi) + \beta(\xi)$$
$$- \lambda^2(N-2)\wp(\xi) + \text{const},$$
$$W(\xi) = \frac{1}{2}[\gamma'(\xi) + \gamma^2(\xi)] + (N-1)\tau(\xi) + \text{const},$$

and the ground-state wave function is

$$\psi_0(x_1, \ldots, x_j + N) = \prod_{j=1}^{N} |\sigma(x_j)|^{\tilde{\lambda}}|\sigma_1(x_j)|^{\tilde{\lambda}_1}|\sigma_2(x_j)|^{\tilde{\lambda}_2}|\sigma_3(x_j)|^{\tilde{\lambda}_3}$$

$$\times \exp\left(\frac{1}{2}\mu_1 x_j^2\right) \prod_{j>k}^{N} |\sigma(x_j - x_k)|^{\lambda}$$

$$\times |\sigma(x_j + x_k)|^{\lambda}\exp[\mu(x_j^2 + x_k^2)],$$

where $\sigma(\xi)$ is the Weierstrass σ function, and

$$\sigma_i(\xi) = \sigma(\xi + \omega_i)\exp[-\xi\zeta(\omega_i)].$$

When the Weierstrass function is degenerate, i.e. $\omega_1 \to \infty$, $\omega_2 \to i\pi/2a$, for $V(\xi)$ and $W(\xi)$ we obtain the following expressions:

$$V(\xi) = \lambda(\lambda - 1)a^2/\sinh^2(a\xi)$$

$$+ \frac{1}{2}\mu\Delta\xi^2 + a\lambda\Delta\xi\coth(a\xi) - 2a^2\lambda\lambda_4\cosh(2a\xi),$$

$$W(\xi) = \lambda_1(\lambda_1 - 1)a^2/\sinh^2(2a\xi) + (\lambda_0 - \lambda_1)$$

$$\times (\lambda_0 + \lambda_1 - 1)a^2/\sinh^2(a\xi)$$

$$+ \left\{\lambda_2\left[2(N-1)\lambda + \lambda_0 + \lambda_1 + \frac{1}{2}\lambda_4 + 1\right]\right.$$

$$\left. - 2\lambda_4(\lambda_0 - \lambda_1)\right\}a^2\cosh(2a\xi)$$

$$+ \left\{\frac{1}{4}\lambda_2^2 - \lambda_4[\lambda_0 + \lambda_1 + 2(N-1)\lambda + 2]\right\}a^2\cosh(4a\xi)$$

$$- \frac{1}{2}\lambda_2\lambda_4\cosh(6a\xi) + \frac{1}{2}\lambda_4^2\cosh(8a\xi) + \frac{1}{2}\lambda_3\Delta\xi^2$$

$$+ \lambda_0\Delta\xi a\coth(a\xi) + \lambda_1\Delta\xi a\tanh(a\xi)$$

$$+ \lambda_2\Delta\xi a\sinh(2a\xi) - \lambda_4\Delta\xi a\sinh(4a\xi), \qquad (2.8)$$

where λ, μ, $\lambda_0, \lambda_1, \lambda_2, \lambda_3, \lambda_4$ are arbitrary constants, and

$$\Delta = \lambda_3 + 2(N-1)\mu.$$

The ground-state wave function then becomes

$$\psi_0(x_1, \ldots, x_N) = \prod_{j>k}^{N} |\sinh[a(x_j - x_k)]|^{\lambda}|\sinh[a(x_j + x_k)]|^{\lambda}$$

$$\times \exp[\mu(x_j^2 + x_k^2)] \prod_{j=1}^{N} |\sinh(ax_j)|^{\lambda_0}$$

$$\times \cosh(ax_j)^{\lambda_1}\exp\left[\frac{1}{2}\lambda_2\cosh(2ax_j)\right.$$

$$\left. + \frac{1}{2}\lambda_3 x_j^2 - \frac{1}{4}\lambda_4\cosh(4ax_j)\right].$$

For the Hamiltonian with potential (2.8) to be self-adjoint, the necessary condition has been found by Meetz

$$\lambda(\lambda - 1) > 3/4.$$

It is obvious that for $\lambda_4 < 0$ the wave function diverges.

When $\Delta = \lambda_4 = 0$, the classical systems with potentials of the binary interaction and external field (2.8) are completely integrable as it was shown by Inozemtsev and Meshcheryakov [40]. When $\lambda_0 = \lambda_1 = \lambda_3 = \lambda_4 = 0$, $\mu = -\beta$, we arrive at the systems considered by Calogero. When $\mu = \Delta = \lambda_0 = \lambda_1$, we obtain the Sutherland–Calogero systems in an external field.

To conclude, we described most general factorization of the wave functions of the systems with potentials (2.8). There is no chance to find excited states. However, for some more simple systems of light particles interacting with particle of infinite mass, the problem of finding discrete spectrum might have a solution which is still not described correctly.

2.3. Equilibrium Points of Classical Integrable Particle Systems, Factorization of Wave Functions of their Quantum Analogs and Polynomial Solutions of the Hill Equation

The most known example of the correspondence between classical and quantum dynamical systems is the Bohr–Sommerfeld rule for one-dimensional quasiclassical motion. It allows one to find the characteristics of the states with large quantum numbers without complicated procedure of investigations of the solutions of the Schrödinger equation. But in general case there is no analogy between the properties of the ground state of the quantum mechanical systems and their classical counterparts.

However, for integrable particle systems there might be a much deeper connection between the classical and quantum dynamics due to common symmetry of the Hamiltonians having the same group theoretical grounds. The aim of this section is demonstration of the existence of such a connection for integrable cases of the motion of the Calogero–Moser particle systems in the external field discovered in Chapter 1. These systems of arbitrary number of particles N with unit mass with the mutual two-particle interaction given by the potential $V(x)$ are supposed to move in the external field created by one particle of infinite mass and described by the potential $W(x)$, and are defined by the quantum Hamiltonian

$$H = \sum_{j=1}^{N} \left[-\left(\frac{\partial}{\partial x_j} \right)^2 + W(x_j) \right] + \sum_{j>k} V(x_j - x_k).$$

For further convenience, we shall slightly change the definition of constants in comparison with Section 2.1, and write the potentials in the form

$$V(x) = g^2 (\sinh x)^{-2},$$

$$W(x) = 8g^2 (2A^2 \cosh 4x + B \cosh 2x + C \sinh 2x). \tag{2.9}$$

The parameters A, B, C and coupling constant g are absolutely arbitrary real numbers. The case of $A = B = C = 0$ corresponds to the usual Calogero–Moser hyperbolic systems. In this notation (differing a bit from the notation in the paper by Inozemtsev and Messhcheryakov), under the condition

$$B = -Ag^{-1}(1 + \alpha(N - 1)), \quad \alpha = \frac{1}{2} + \sqrt{g^2 + \frac{1}{4}} \tag{2.10}$$

the ground-state wave function of the quantum systems defined by (2.9) is factorized,

$$\psi(x_1, \ldots, x_N) = \prod_{j>k}^{N} |\sinh(x_j - x_k)|^{\alpha}$$

$$\times \prod_{j=1}^{N} \exp \left[-g \left(4A \cosh 2x_j + \frac{C}{A} x_j \right) \right]. \tag{2.11}$$

Later on, it was shown by Sasaki and Takasaki [31] that the form of the Hamiltonian can be enlarged in such a way that the model becomes one of the class of quasiexactly solvable ones, i.e. it is possible to determine analytically some finite set of the eigenvalues of the Hamiltonian. It was found also that the eigenvalue which corresponds to the wave function (2.11), i.e. the ground-state energy, is given by the formula

$$E_q = -\frac{N}{2}\left[\frac{\alpha^2}{3}(N^2-1) + \frac{C^2}{A^2}g^2 - 32A^2g^2\right]. \qquad (2.12)$$

We use the system of units in which $h = 1$; the parameters of the potentials (2.9) slightly differ from that used in the paper by Inozemtsev and Meshcheryakov [40] for convenience. If the condition (2.10) does not take place, the factorization (2.11) does not hold and the energy of the ground state depends on the parameters A, B, C through much more complicated forms. The solution of the corresponding quantum problem is not known to date.

The natural question arises: which properties of the corresponding *classical* systems do correspond to (2.10)–(2.12)? It is natural to suppose that they reveal for the states with minimal energy, i.e. classical equilibrium points. The description of these points might be much easier under the condition similar to (2.10). It will be shown later that such a correspondence indeed takes place. Note also that at $A = B = C = 0$, there are no equilibrium points at all.

It is well known that there is a relation between the coordinates of particles of some, more simple than (2.10), classical systems of particles at equilibrium, and zeroes of classical orthogonal polynomials (for example, see Section 1.5 of Chapter 1). It was discovered first by Stiltjes in the 19th century. It was used for getting various "sum rules" for zeroes of Hermite, Jacobi, Laguerre polynomials and zeroes of Bessel functions. As for the systems (2.9) in some limit, the relation when $W(q)$ coincides with the Morse potential $8g^2A^2(\exp(4x) - 2\exp(2x))$ was established in Section 1.5 of Chapter 1. Let us now show that the simplification of the equilibrium equations takes place even in the general case of potentials (2.9). These equations can be written as

$$-\sum_{k\neq j}\frac{\cosh(x_j - x_k)}{\sinh^3(x_j - x_k)}$$

$$+ 8[4A^2\sinh 4x_j + B\sinh 2x_j + c\cosh 2x_j] = 0.$$

With the use of variables $z_j = \exp(2x_j)$, this system of equations can be cast in the rational form

$$-\sum_{k \neq j} \frac{z_k(z_j + z_k)}{(z_j + z_k)^3} + 4A^2(z_j - z_j^{-3}) + B + C - (B - C)z_j^{-2} = 0.$$

$$(2.13)$$

Let us now use the following trick introduced by Ahmed [42] for transformation of the system (2.13) to a more simple form. Let $\{z_j\}$ be the solution of (2.13). Let us construct the polynomial

$$PN(z) = \prod_{j=1}^{N}(z - z_j),$$

and consider the integral on a closed contour removed enough from the origin and surrounding all $\{z_j\}$, for the function

$$F_j(z) = \frac{z(z + z_j)P_N'(z)}{(z - z_j)^3 P_N(z)}.$$

Since $F_j(z) \sim z^{-2}$ as $z \to \infty$, this integral and, correspondingly, the sum of the residues at the poles of $F_j(z)$, equal zero. The residues at the poles of the first order at $z_k, k \neq j$, equal

$$\mathrm{res}F_j(z)|_{z=z_k} = \frac{z_k(z_k + z_j)}{(z_k - z_j)^3},$$

and the system (2.13) can be written as

$$4A^2(z_j - z_j^{-3}) + B + C - (B - C)z_j^{-2} = -\mathrm{res}F_j(z)|_{z=z_j}. \quad (2.14)$$

The pole of $F_j(z)$ at the point $z = z_j$ is of the fourth order, and the right-hand side of the above equation is calculated with the use of the formula

$$-\mathrm{res}F_j(z)|_{z=z_j} = -\frac{1}{6}\frac{d^3}{d\xi^3}\left[\xi(z_j + \xi)(2z_j + \xi)\frac{P_N'(z_j + \xi)}{P_N(z_j + \xi)}\right]\Big|_{\xi=0}$$

$$= -[(a + 3z_j(b - a^2) + z_j^2(2a^3 - 3ab + c)],$$

where

$$a = \frac{P_N''(z_j)}{P_N'(z_j)}, \quad b = \frac{P_N^{(3)}(z_j)}{P_N'(z_j)}, \quad c = \frac{P_N^{(4)}(z_j)}{P_N'(z_j)}.$$

If we suppose that $P_N(z)$ is a solution of the second-order differential equation

$$\mu(z)P_N''(z) + \rho(z)P_N'(z) + \lambda(z)P_N(z) = 0, \qquad (2.15)$$

(the functions μ, ρ, λ can depend on N), then the values a, b, c and the residue res $F_j(z)|_{z=z_j}$ can be expressed through μ, ρ, λ and their derivatives. Our purpose is to choose such kind of μ, ρ, λ which guarantee the transformation of the Equation (2.14) to identities, and also allow one to prove the existence of the polynomial solution to Equation (2.15). One can show that all these requirements can be satisfied if and only if

$$\mu(z) = z^2.$$

Then the residue can be relatively simply expressed through ρ and λ,

$$\operatorname{res} F_j(z)|_{z=z_j} = \frac{1}{4}\left(-\frac{\rho^2}{z_j^3} + \frac{\rho\rho'}{z_j^2} - \rho'' - 2\lambda'\right).$$

The validity of Equation (2.14) happens if

$$\rho(z) = pz^2 + qz + r, \quad \lambda(z) = uz + v.$$

The constants p, q, r, u are determined from (2.15),

$$p = -r = -4A, \quad pq - 2(p + u) = 4(B + C), qr = 4(B - C).$$
$$(2.16a)$$

The polynomial of the degree N can be a solution of Equation (2.15) in accordance with the choice of $\rho(z)$ and $\lambda(z)$ if and only if the condition $pN + u = 0$ is satisfied. It results finally in

$$B = -A(N - 1),$$

$$q = -\frac{C}{A} - N + 1. \qquad (2.16b)$$

It is clear now that the solutions to the system (2.13) are determined by the roots of the polynomial $P_N(z)$ satisfying the second-order differential equation (2.15) if and only if the above equalities take place. The coefficient B cannot be arbitrary. Equation (2.15) with the coefficients quadratic in z does not fall into the hypergeometric

class since it has irregular singular points at zero and infinity. This is
the characteristic feature of the Hill equation. The parameter v must
be determined by the condition of compatibility of the system of the
recurrence relations for the coefficients of the polynomial

$$P_N(z) = \sum_{l=0}^{N-1} d_{N-l} z^l + z^N, \qquad (2.16c)$$

which appear under the substitution of it into (2.15). This condition
can be represented in the form of the algebraic equation of $(N+1)$th
order. Its solution for v should be chosen so as all the roots of
$P_N(z)$ must be real and positive. We cannot point out the explicit
dependence of v on A and C. However, it turns out that it is
not necessary for the calculation of the minimum energy at the
equilibrium.

   This calculation can be performed as follows. Let us write the
expression for the minimum energy at the equilibrium in the form

$$E_{cl} = 4g^2(S + 2A^2(S_2 + S_{-2}) + (B+C)S_1 + (B-C)S_{-1},$$

$$(2.17a)$$

where

$$S = \sum_{j>k} \frac{z_j z_k}{(z_j - z_k)^2},$$

$$S\alpha = \sum_{j=1}^{N} z_j^\alpha.$$

The sum of the residues of the function

$$\psi(z) = \sum_{j=1}^{N} \frac{z z_j}{(z - z_j)^2} \frac{P'_N(z)}{P_N(z)}$$

in all its poles must equal zero since $\psi(z) \sim z^{-2}$ as $z \to \infty$.
The calculation of these residues can be done within the scheme
used earlier for calculation of $\{z_j\}$. The equation (2.15) with the
coefficients (2.16a) and (2.16b) allows us to express the double sum

through S_α,

$$S = -\frac{p^2}{24}(S_2 + S_{-2}) - \frac{S_1}{12}(pq - 3p - 2a) - \frac{S_{-1}}{12}r(q+1)$$
$$-\frac{N}{12}(2pr + (q-1)^2 - 4v - 1). \tag{2.17b}$$

All the sums $\{S_\alpha\}$ are expressed through p, q, r, u, v with the use of (2.15) and (2.16c). It follows from (2.16c) that

$$S_1 = -d_1, \quad S_1^2 - S_2 = 2d_2,$$

$$S_{-1} = -\frac{d_{N-1}}{d_N}, \quad S_{-1}^2 - S_{-2} = 2\frac{d_{N-2}}{d_N}.$$

The substitution of (2.16c) into (2.15) leads to simple expressions for all values in the right-hand sides of all these equations. The subsequent substitution of them into (2.17a) and (2.17b) results, after very long but not too tedious calculations, in the very simple formula

$$E_{cl} = \frac{Ng^2}{2}\left[\frac{N^2-1}{3} + \frac{C^2}{A^2} - 32A^2\right].$$

All non-zero degrees of the parameter v which appear on intermediate stages of the calculation cancel, and getting the final answer does not need the explicit form of v.

Let us show also that the Equation (2.15) with the coefficients (2.16a) and (2.16b) can be considered as a one-particle Schrödinger equation. Indeed, after the change of the variable $z = \exp 2x$ and the substitution

$$P_N(e^{2x}) = \psi(x)\exp\left[4A\cosh 2x + \left(\frac{C}{A} + N\right)x\right],$$

one obtains from (2.15) the equation

$$-\frac{\psi''}{2} + \tilde{W}(x)\psi = \varepsilon\psi,$$

where

$$\tilde{W}(x) = 8(2A^2 \cosh 4x - A(N+1) \cosh 2x + C \sinh 2x),$$

$$\varepsilon = 16A^2 + 2v - \frac{1}{2}\left(\frac{C}{A} + N\right)^2.$$

To summarize, we establish that under the condition (2.16b) the coordinates of particles of the classical systems with potentials (2.9) at equilibrium coincide with zeroes of the wave functions of Nth level of one-particle wave function, the solution to the one-dimensional Schrödinger equation with potential $\tilde{W}(x)$. The first of conditions (2.16b) is completely analogous to the condition of the factorization of the wave function of the ground state for the *quantum* Calogero–Moser problem (2.11) and exactly coincides with it in the limit of large coupling constants g (i.e. at $\hbar \to 0$). In this limit, the minimal classical and quantum energies do coincide; the first of condition (2.16b) and condition for the factorization of the wave function of the Calogero–Moser system in potential of external field (2.9) also coincide. The dependence of classical equilibrium state and ground-state quantum energies in this limit are identical. It seems that this fact is not casual. Nevertheless, we do not know the theoretical group interpretation of such a coincidence. It is quite possible that it might be found if one finds a way to include (2.9) into the Kirillov–Kostant scheme of geometrical quantization. However, till now it was used only for the investigation of much more simple dynamical systems.

2.4. The Structure of Eigenvectors of the Multidimensional Lamé Operator

The problem of finding explicit solutions to quantum Calogero–Moser systems with elliptic potential of mutual interaction seemed for a long time to be completely unsolved. It can be written as an eigenvalue problem for eigenvectors of the Schrödinger operator

$$H_{N,n} = -\frac{1}{2}\sum_{j=1}^{N}\left(\frac{\partial}{\partial x_j}\right)^2 + n(n+1)$$

$$\times \sum_{j>l}\wp(x_j - x_l), \quad n \in \mathbf{Z}_+, \tag{2.18}$$

where $\wp(x)$ is the Weierstrass function with two periods ω_1 and ω_2, $\Im m(\omega_2/\omega_1) > 0$. In physical application, where $H_{N,n}$ has to be Hermitian, these periods are usually chosen such that $\omega_1 \in \mathbf{R}_+$, $i^{-1}\omega_2 \in \mathbf{R}_+$. We shall describe here the results of Dittrich and Inozemtsev [44].

The above elliptic operator describes the quantum system of N particles interacting via two-body potential $n(n+1)\wp(x)$. The analogous classical many-particle problem has been proven to be completely integrable in the Liouville sense by Perelomov and solved in terms of N-dimensional Riemann theta functions by Krichever [43]. The construction of the set of corresponding quantal "integrals of motion" $\{I_l\}(1 \leq l \leq N-1)$ commuting with $H_{N,n}$, has also been performed by Olshanetsky and Perelomov by using the quantum analogue of the classical Lax representation. Moreover, the superintegrability for integer n, i.e. existence of more than $N-1$ functionally independent operators commuting with $H_{N,n}$, has been conjectured by Chalykh and Veselov [45] and proved in the simplest case $N = 2$ in which the corresponding eigenproblem reduces to the usual Lamé equation and the structure of eigenfunctions was described more than a century ago by Hermite. Note that the Hermite solution contains only Weierstrass sigma functions. Despite many results obtained by Sutherland, Sekiguchi, Chalykh and Veselov for trigonometric and hyperbolic degenerations of (2.18), the explicit form of any solution to $H_{N,n}\psi = E\psi$ for $N \geq 3$ was an enigma for a long time. Later, we shall describe its solution found by Dittrich and Inozemtsev. But till now it is not clear how one should use the rather complicated operators $\{I_l\}$ containing higher derivatives for reduction of the spectral problem.

Our observation consists in using the simpler symmetry of HN, n for this purpose. Since $\wp(x)$ is double periodic, it is easy to see that (2.18) commutes with the $2N$ shift operators

$$Q_{\alpha j} = \exp\left(\omega_\alpha \frac{\partial}{\partial x_j}\right), \quad \alpha = 1, 2, \ 1 \leq j \leq N.$$

Let $\psi(q)(x_1, \ldots, x_N)$ be their common eigenvector,

$$\psi^{(q)}\left(x_1 + \sum_{\alpha=1}^{2} l_1^{(\alpha)}\omega_\alpha, \ldots, x_N + \sum_{\alpha=1}^{2} l_N^{(\alpha)}\omega_\alpha\right)$$

$$= \exp\left(i\sum_{j=1}^{N}\sum_{\alpha=1}^{2} q_\alpha^{(j)} l_j^{(\alpha)}\right)\psi^{(q)}(x_1, \ldots, x_N),$$

where $l_j^{(\alpha)} \in \mathbf{Z}$ and $q_\alpha^{(j)} \in \mathbf{C}(\mathrm{mod}2\pi)$. Hence, $\psi^{(q)}(x_1, \ldots, x_N)$ can be treated on the N-dimensional torus

$$\mathbf{T}^N = (\mathbf{C}/\Gamma)^N, \ \Gamma = \mathbf{Z}\omega_1 + \mathbf{Z}\omega_2$$

with quasiperiodic boundary conditions. The structure of singularities of $H_{N,n}$ on this torus shows that $\psi^{(q)}(x_1, \ldots, x_N)$ is analytic on $\mathbf{T}^N \backslash \mathbf{P}_N$, where $\mathbf{P}_N$ is the set which consists of all $N(N-1)/2$ hypersurfaces P_{jk} defined by the equalities $x_j = x_k$, $1 \leq j < k \leq N$. On each P_{jk}, since n is an integer, $\psi^{(q)}(x_1, \ldots, x_N)$ has a pole of nth order

$$\psi^{(q)}(x_1, \ldots, x_N)|_{x_j \to x_k} \sim (x_j - x_k)^{-n}\phi^{(jk)}(x_1, \ldots, x_N),$$

where $\phi(jk)$ do not contain any singularities as $x_j \to x_k$. Let $\Psi_{N,n}$ be the class of functions analytic on $\mathbf{T}^N \backslash \mathbf{P}_N$ satisfying the above relations.

The main result consists in combining these properties so as to reduce the problem of finding $\psi^{(q)}$ to an algebraic one.

Note that the Weierstrass sigma function has only one zero on the torus $\mathbf{T} = \mathbf{C}/\Gamma$,

$$\sigma(x) = x \prod_{m^2+n^2 \neq 0}^{\infty} \left(1 - \frac{x}{m\omega_1 + n\omega_2}\right)$$

$$\times \exp\left[\frac{x}{m\omega_1 + n\omega_2} + \frac{1}{2}\left(\frac{x}{m\omega_1 + n\omega_2}\right)^2\right].$$

Hence, all singularities of $\psi^{(q)}$ may be written in the form

$$\varphi(x_1, \ldots, x_n) = \prod_{j>k}^{N}(\sigma(x_j - x_k))^{-n},$$

and the function

$$\xi(x_1, \ldots, x_N) = \varphi(x_1, \ldots, x_N)\psi^{(q)}$$

is entire on the torus **T**.

Proposition 1. *The class* $\Psi_{N,n}$ *is a functional manifold of dimension* $2N - 1 + n(Nn)^{N-2}$. *The parameters* $\{q_\alpha^{(j)}\}$ *are not independent but connected by the linear relation*

$$\sum_{j=1}^{N}(q_1^{(j)}\omega_2 - q_2^{(j)}\omega_1) \in 2\pi\Gamma.$$

The manifold $\Psi_{N,n}$ can be described as a union of the $(2N-1)$-parametric family of linear spaces $L(q)$, dim $L(q) = n(nN)^{N-2}$, with the basic vectors parametrized by $q = \{q_\alpha^{(j)}\}$ satisfying the previous equation.

The scheme of the proof is based on investigating the entire functions $\{\xi\}$ arising after explicit representation of the leading singularities of $\psi^{(q)}(x_1, \ldots, x_N)$ in terms of inverse powers of the Weierstrass sigma functions. After selecting the factor of quasiperiodicity, the functions $\xi(x_1, \ldots, x_N)$ can be expanded into Fourier series, coefficients of which are determined by the second condition of quasiperiodicity up to some big amount of constants.

Proposition 2. *The co-ordinate system on* $\Psi_{N,n}$ *can be chosen in such a way that all its elements are expressed through the Riemann theta function of genus 1.*

The first part of the proof consists in the explicit representation of the basis vectors of $L(q)$ in terms of the $(N-1)$-dimensional theta functions with the Riemannian B matrix having all diagonal and all non-diagonal elements equal,

$$B_{jk} = (\pi i\omega_2/nN\omega_1)(1 + \delta_{jk}), \quad 1 \le j, k \le N - 1.$$

The second part is simply an application of the Appell [46] reduction theorem. Since the Riemann theta functions of genus 1 are expressed through the Weierstrass sigma function, we shall work with the last one.

Let $\Phi_{N,n}$ be the submanifold of $\Psi_{N,n}$ which consists of the eigenvectors of $H_{N,n}$. The elements of $\Phi_{N,n}$ are determined by $N + 1$ complex parameters which include the $N - 1$ relative particle quasimomenta, the total momentum and the trivial normalization factor. So the problem of selecting $\Phi_{N,n}$ from $\Psi_{N,n}$ is equivalent to finding the solutions to $N - 2 + n(Nn)^{N-2}$ purely algebraic equations which arise under substitution of the general expression for the elements of $\Psi_{N,n}$ into the eigenequation $H_{N,n}\psi = E\psi$.

Two years later, Felder and Varchenko [47] presented the formula for arbitrary N and n. It was obtained at the investigation of the asymptotic of the solutions of the Knizhnik–Zamolodchikov–Bernard equation. However, it is very cumbersome and excludes any practical calculations.

2.5. The Discrete Spectrum of New Exactly Solvable Quantum N-body Problem on a Line

In this section, we shall study the discrete spectrum of the quantum problem with Hamiltonian

$$H = \sum_{j=1}^{N} \left[\frac{p_j^2}{2} + 2A^2(\exp(4x_j) - 2(\exp(2x_j)) \right]$$
$$+ \sum_{j>k}^{N} \alpha(\alpha - 1)\sinh^{-2}(x_j - x_k), \qquad (2.19)$$

i.e. hyperbolic Sutherland systems interacting with one particle of infinite mass creating the external field with the Morse potential [48]. For the Hamiltonian (2.19) to be self-adjoint, it is necessary to constrain constants α and A in (2.21): $\Im A = \Im \alpha = 0$, $\Re \alpha \geq 3/2$. We shall follow the paper by Inozemtsev and Meshcheryakov [48].

The Hamiltonian (2.19), evidently, has also the states with continuous spectrum corresponding to the scattering of M particles in bound states of $(N - M)$ ones, $N > M$. These processes will not be discussed here; we are interested only in the non-trivial discrete spectrum of the Hamiltonian (2.19).

The ground-state wave function for the Schrödinger operator (2.19) was found by us earlier (Equation (2.3)). Note that variables in the corresponding Schrödinger equation are not separable even in the simplest case $N = 2$. One can also construct non-trivial quantum integrals of motion, the operators containing higher degrees of momenta commuting with each other and with the Hamiltonian.

Performing in (2.19) the change of variables $z_j = \exp(2x_j)$ $(0 \leq z_j < \infty)$, we transform the Schrödinger equation corresponding to the Hamiltonian (2.19) as follows:

$$\left\{ \sum_{j=1}^{N} \left[-2\left(z_j^2 \frac{\partial^2}{\partial z_j^2} + z_j \frac{\partial}{\partial z_j} \right) + 2A^2(z_j^2 - 2z_j) \right] \right.$$

$$\left. + \sum_{j>k}^{N} \frac{4\alpha(\alpha - 1)z_j z_k}{(z_j - z_k)^2} \right\} \psi = E\psi. \tag{2.20}$$

Further, for brevity we shall suppose the particles obey Bose statistics. For the reduction of powers of singularities in Equation (2.20), let us use a standard trick and introduce a function $\Phi(z_1, \ldots, z_N)$ by the relation

$$\psi(z_1, \ldots, z_N) = \left(\prod_{j>k}^{N} |z_j - z_k|^\gamma \prod_{s=1}^{N} z_s^\beta \exp(-\tau z_s) \right) \Phi(z_1, \ldots, z_N). \tag{2.21}$$

Let us choose the constants γ and τ equal to α and A $(A > 0)$, respectively. The leading singularities in Equation (2.22) are cancelled out and the equation for function $\Phi(z_1, \ldots, z_N)$ can be represented in the form

$$-2 \sum_{j=1}^{N} \left[z_j^2 \frac{\partial^2 \Phi}{\partial z_j^2} + \left(z_j(1 + 2\beta) - 2Az_j^2 + 2\alpha \sum_{j \neq k, j, k=1}^{N} \frac{z_j^2}{z_j - z_k} \right) \frac{\partial \Phi}{\partial z_j} \right]$$

$$+ \left[p \sum_{j=1}^{N} z_j - (E - E(\beta)) \right] \Phi = 0, \tag{2.22}$$

where

$$p(\beta) = 2A(2\alpha(N-1) + 1 + 2\beta - 2A),$$

$$E(\beta) = -N\left[2\beta^2 + 2\alpha\beta(N-1) + \tfrac{\alpha^2}{3}(N-1)(2N-1)\right]. \quad (2.23)$$

According to Bose statistics $\Phi(z_1, \ldots, z_N)$ must be symmetric under the interchange of each of the two arguments. Let us represent it as a function of N symmetric polynomials $a_1, \ldots, a_N$

$$\Phi(z_1, \ldots, z_N) = \Phi(a_1, \ldots, a_N),$$

$$a_1 = \prod_{j=1}^{N} z_j, \quad a_l = \frac{\hat{D}^{l-1}}{(l-1)!} a_1, \quad l = 1, \ldots, N, \quad \hat{D} = \sum_{j=1}^{N} \frac{\partial}{\partial z_j}, \quad (2.24)$$

In what follows, where the indices of quantities $\{a_s\}$ exceed N or are less than 1, one must put $a_{N+1} = 1$, $a_0 = a_{N+2} = a_{N+3} = \cdots = 0$.

By simple calculations taking into account the properties of polynomial a_1 and operator $\hat{D}$, we obtain the following relations:

$$\sum_{j=1}^{N} z_j \frac{\partial a_l}{\partial z_j} = (N - l + 1)a_l, \quad (2.25a)$$

$$\sum_{j=1}^{N} z_j^2 \frac{\partial a_l}{\partial z_j} = a_N a_l + a_{l-1}(l - N - 2) \quad (2.25b)$$

$$\hat{C}_2 a l = \sum_{j \neq k, j, k = 1}^{N} \frac{z_j^2}{z_j - z_k} \frac{\partial a_l}{\partial z_j} = \frac{1}{2}(N - l + 1)(N + l - 2)a_l. \quad (2.25c)$$

The proof of the last relation is more complicated. Let us introduce an equivalent but a bit more simple representation of the polynomials $\{a_l\}$:

$$a_l = \frac{1}{(l-1)!} \left(\frac{d}{d\lambda}\right)^{l-1} \prod_{s=1}^{N} (z_s + \lambda)|_{\lambda=0}.$$

Now,

$$\hat{C}_2 a_l = \frac{1}{2} \sum_{j>k,k=1}^{N} \left(\frac{z_j^2}{z_j - z_k} \frac{\partial a_l}{\partial Z_j} - \frac{z_k^2}{z_j - z_k} \frac{\partial a_l}{\partial z_k} \right).$$

But,

$$\frac{\partial a_l}{\partial z_j} = \frac{l}{(l-1)!} \left(\frac{d}{d\lambda} \right)^{l-1} \frac{\partial}{\partial z_j} \prod_{s=1}^{l} (z_s + \lambda)|_{\lambda=0}$$

$$= \frac{l}{(l-1)!} \left(\frac{d}{d\lambda} \right)^{l-1} \frac{1}{z_j + \lambda} \prod_{s=1}^{N} (z_s + \lambda)|_{\lambda=0}.$$

Hence,

$$\hat{C}_2 a_l = \frac{1}{2} \sum_{j>k,k=1}^{N} \frac{1}{(l-1)!} \left(\frac{d}{d\lambda} \right)^{l-1} \prod_{s=1}^{N} (z_s + \lambda)$$

$$\times \left[\frac{z_j^2}{(z_j + \lambda)} - \frac{z_k^2}{(z_k + \lambda)} \right] \frac{1}{z_j - z_k} |_{\lambda=0}.$$

But,

$$\frac{z_j^2}{z_j + \lambda} - \frac{z_k^2}{z_k + \lambda} = \frac{(z_j - z_k)[(z_j + \lambda)(z_k + \lambda) - \lambda^2]}{(z_j + \lambda)(z_k + \lambda)}.$$

Now, we can see that the singularity $(z_j - z_k)$ cancels and we can write the following representation for $\hat{C}_2 a_l$:

$$\hat{C}_2 a_l = \frac{1}{2} \sum_{j>k,k=1}^{N} \frac{1}{(l-1)!} \left(\frac{d}{d\lambda} \right)^{l-1}$$

$$\times \left[\prod_{s=1}^{N} (z_s + \lambda) - \lambda^2 \prod_{s=1, s \neq j,k} (z_s + \lambda) \right] |_{\lambda=0}.$$

It is easy to see that the first term in the bracket just equals $\frac{N(N-1)}{2}a_l$. As for the second term,

$$\frac{1}{(l-1)!}\left(\frac{d}{d\lambda}\right)^{l-1}\left[\lambda^2\prod_{s=1,s\neq j,k}(z_s+\lambda)|_{\lambda=0}\right]$$

$$=\frac{(l-1)(l-2)}{(l-1)!}\left(\frac{d}{d\lambda}\right)^{l-3}\prod_{s=1,s\neq j,k}^{N}(z_s+\lambda)|_{\lambda=0},$$

it is a homogeneous polynomial of the degree $N-l-3$ of $N-2$ variables. It contains $\frac{(N-2)!}{(N-l-1)!(l-3)!}$ terms. The sum $\sum_{j>k,k=1}\frac{1}{(l-3)!}\left(\frac{d}{d\lambda}\right)^{l-3}\prod_{s=1,s\neq j,k}(z_s+\lambda)|_{\lambda=0}$ contains

$$\frac{\frac{N(N-1)}{2}\times(N-2)!}{(N-l-1)!(l-3)!}=\frac{1/2(l-1)(l-2)N!}{(N-l-1)!(l-1)!}$$

terms. The $\frac{N!}{(N-l-1)!(l-1)!}$ terms form the polynomial of N variables of the degree $N-l+1$, i.e. a_l. Hence,

$$\hat{C}_2 a_l=\frac{1}{2}[N(N-1)-(l-1)(l-2)]a_l$$

$$=\frac{1}{2}(N-l+1)(N+l-2)a_l.$$

Q.E.D.

By easy calculation, one can also find the formula

$$\frac{\sum_{j=1}^{N}z_j^2\partial a_l}{\partial z_j}\times\partial a_m \frac{}{\partial z_j}=(N-l+1)a_l a_m-\sum_{r=1}^{l}(l+m-2r)a_r a_{l+m-r}.$$

(2.25d)

By (2.25a)–(2.25d), it is easy to find the equation for function $\Phi(a_1,\ldots,a_N)$ (2.25a):

$$(\hat{H}_1+\hat{H}_2)\Phi(a_1,\ldots,a_N)=(E-E(\beta))\Phi(a_1,\ldots,a_N),\qquad(2.26)$$

where

$$\hat{H}_1 = -2 \left\{ \sum_{l,m=1} \left[(N-l+1)a_l a_m - \sum_{r=1}^{l} (l+m-2r)a_r a_{l+m-r} \right] \right.$$

$$\times \frac{\times \partial^2}{\partial a_l \partial a_m} + \sum_{l=1}^{N} \left[(N-l+1)(1+2\beta + \alpha(N+l-2))a_l \frac{\partial}{\partial a_l} \right.$$

$$\left. \left. + 2A(N-l+2)a_{l-1}\frac{\partial}{\partial a_l} \right] \right\} \tag{2.27}$$

$$\hat{H}_2 = a_N \left(4A \sum_{l=1}^{N} a_l \frac{\partial}{\partial a_l} + p(\beta) \right). \tag{2.28}$$

Let us choose the solutions of Equation (2.26) as polynomials in variables $\{a_l\}$ of the form

$$\Phi^{(n)}(a_1, \ldots, a_N) = \sum_{\mu=0}^{n} \sum_{j_1+\cdots+j_N=\mu, 0 \le j \le \mu} c_{j_1 \cdots j_N}^{(\mu)} a_1^{j_1} \ldots a_N^{j_N}. \tag{2.29}$$

Evidently the operator $\hat{H}_1$ does not raise the maximal degree of these polynomials. As for $\hat{H}_2$, its adds 1 to this degree if the condition

$$p(\beta) + 4An = 0 \tag{2.30}$$

is not satisfied.

If $p(\beta)$ obeys Equation (2.30), the sum $\hat{H}_1 + \hat{H}_2$ is a linear operator acting in the space of polynomials in N variables with degree not exceeding n. The dimensionality of this space is

$$\frac{(N+n)!}{n!N!}.$$

The eigenvalues of this operator represent (at given n) the spectrum of the considered problem up to a constant $E(\beta)$. In this case, the parameter β and the constant $E(\beta)$ are completely determined by the integer n.

The normalizability condition for the wave function (2.21) has the form

$$\int dz_1, \ldots, dz_N \prod_{j>k}^{N} |z_j - z_k|^{2\alpha} \prod_{s=1}^{N} z_s^{2\beta-1}$$

$$\times \exp(-2Az_s)|\Phi(z_1, \ldots, z_N)|^2 < \infty. \qquad (2.31)$$

From (2.29) and (2.31) it follows that $\beta(n)$ must obey the condition $\beta > 0$, hence the integer n is restricted:

$$A - \alpha(N - 1) > n + 1/2. \qquad (2.32)$$

So, we formulate the way for constructing the discrete spectrum of the Hamiltonian (2.19): for the integers n which obey the condition (2.32), one must find the matrix of the operator $\hat{H}_1 + \hat{H}_2$ according to (2.26)–(2.28) in the basis

$$\{a_1^{j_1}, \ldots, a_N^{j_N}\}, \quad j_1+, \ldots, j_N < n.$$

The eigenvalues of this matrix, up to constant $E(\beta_n)$, represent the spectrum to be found. So, the problem is reduced to an algebraic equation. This problem is apparently simple in the case $n = 1$, where $\Phi(a_1, \ldots, a_N)$ has the form

$$\Phi^{(1)}(a_1, \ldots, a_N) = \sum_{l=1}^{N} c_l a_l + c_{l+1}, \qquad (2.33)$$

The operator $\partial^2/\partial a_l \partial a_m$ acting on the function (2.33) reduces it to zero. The matrix of the operator $\hat{H}_1 + \hat{H}_2$ is the upper triangular one and its spectrum is determined by the diagonal elements:

$$-E_l^{(1)} = \frac{N}{2} \left[(2A - \alpha(N - 1) - 3)^2 + \frac{\alpha^2(N^2 - 1)}{3} \right]$$

$$+ 2(N - l + 1)(2A - 2 - \alpha(N - l)), \quad l = 2, \ldots, N + 1. \qquad (2.34)$$

Note that at $3/2 < A - \alpha(N - 1) < 5/2$ the values $E_l^{(1)}$ (2.34) and the ground-state energy represent the whole discrete spectrum of the problem we are interested in. One can also (up to normalization constants) determine the wave functions corresponding to eigenvalues $E_l^{(1)}$:

$$\Phi_l^{(1)}(a_1, \ldots, a_N) = \sum_{j=1}^{l} (-1)^{l-j} \left(\frac{2A}{\alpha}\right)^{l-j} \frac{(N-j+1)!}{(l-j)!(N-l+1)!}$$

$$\times \frac{\Gamma\left(\frac{2A-2}{\alpha} - 2N + j + l - 1\right)}{\Gamma\left(\frac{2A-2}{\alpha} - 2(N-l) - 1\right)} a_j,$$

where Γ is the Euler gamma function.

When $n > 1$, it is easy to see the matrix of the operator $\hat{H}_1$ in the basis $\{a_1^{j1}, \ldots, a_N^{jN}\}$ is no longer upper triangular, and the determination of eigenvalues represents a much more complicated problem. Possibly, for the solution of this problem one can apply algebraic methods as followed by Olshanetsky and Perelomov.

We can, however, immediately find one of these eigenvalues. Really, note that $\hat{H}_1 + \hat{H}_2$ transforms (under the condition (2.30)) the linear space of polynomials of the form (2.29) to a space of a lower dimensionality: its action on $a_1^{j1}, \ldots, a_N^{jN}$ does not contain the polynomial of zero degree. So, there exists a subspace corresponding to the zero eigenvalue of this operator and

$$E(N) = E(\beta_n) = -\frac{N}{2} \left[\frac{\alpha}{3}(N^2 - 1) + (2A - \alpha(N-1) - 2n - 1)^2\right]$$

are eigenvalues of the Hamiltonian. It is evident that the set $\{E^{(n)}\}$, up to a constant $-\frac{N\alpha^2}{6}(N^2 - 1)$, coincides with the set of the levels of unperturbed system of N particles in the Morse oscillator, for which each particle is on an n-th level and the constant A is "renormalized" by the interaction $A \to A - \frac{\alpha}{2}(N-1)$.

2.6. Wave Functions of Discrete Spectrum States of Integrable Quantum Systems with N Degrees of Freedom

In this section, we shall describe the wave functions of the discrete spectrum of the Hamiltonian

$$H = \sum_{j=1}^{N} \frac{1}{2} p^2 j + \sum_{j>k}^{N} [V(x_j - x_k) + V(x_j + x_k)] + \sum_{j=1}^{N} W(x_j),$$

where

$$V(x) = \frac{\alpha(\alpha-1)}{\sinh^2(x)},$$

$$W(x) = \frac{g}{\cosh^2(x)}, \quad g = -\frac{1}{2}\lambda(\lambda+1) \quad \lambda > 0. \tag{2.35}$$

Note that for $g > 0$ there may also exist only a finite number of states of the discrete spectrum as for the one-dimensional problem of motion of a single particle in the potential $W(x)$ (2.35). In the following, we shall show how this number is determined by the values of the constants λ and α (for the Hamiltonian to be hermitian, the inequality $\alpha(\alpha-1) \geq 3/4$, i.e. $\alpha \geq 3/2$ should hold).

At $\alpha = 0$ the problem is trivial: the Hamiltonian (2.5) describes a set of N non-interacting particles in the external field

$$-\frac{1}{2}\lambda(\lambda+1)\cosh^{-2}(x).$$

The Schrödinger equation reduces in this case to the hypergeometrical equation by the substitution $z = \tanh(x)$.

Making the change $z_j = \tanh(x_j), j = 1, \ldots, N$, in the case of the non-trivial interaction ($\alpha \neq 0$) we are interested in, we arrive at the equation [49]

$$\left[\sum_{j=1}^{N} \left(\frac{1}{2}(1-z_j^2)\frac{\partial}{\partial z_j}(1-z_j^2)\frac{\partial}{\partial z_j} + \lambda(\lambda+1)(1-z_j^2) \right) \right.$$

$$\left. + \sum_{j>k,k=1}^{N} \frac{2\alpha(\alpha-1)(1-z_j^2)}{(z_j^2-z_k^2)^2}(1-z_k^2)(z_j^2+z_k^2) \right] \psi = E\psi. \tag{2.36}$$

Owing to the symmetry of the problem, it is sufficient to consider the motion in a polyhedral angle $\{z_j \pm z_k \geq 0, j > k\}$. To simplify Equation (2.36), we introduce the new function

$$\psi(z_1, \ldots, z_N) = \varphi(z_1, \ldots, z_N)\chi_{\alpha,\beta}(z_1, \ldots, z_N), \tag{2.37}$$

where $\chi_{\alpha,\beta}(z_1, \ldots, z_N)$ is an analog of the ground-state wave function of the system under consideration,

$$\chi_{\alpha,\beta}(z_1, \ldots, z_N) = \prod_{j=1}^{N}(1-z_j^2)^\beta \prod_{j>k,k=1}^{N}(z_j^2-z_k^2)^\alpha. \tag{2.38}$$

Then the equation for φ becomes

$$\tilde{H}\varphi = [E - E(\beta)]\varphi, \tag{2.39a}$$

where

$$\tilde{H} = -\frac{1}{2}\left\{ \sum_{j=1}^{N}\left[(1 - z_j^2)^2 \frac{\partial^2}{\partial z_j^2} \right.\right.$$

$$+ \left(4\alpha z_j(1 - z_j^2)^2 \sum_{k \neq j, j=1}^{N} (z_j^2 - z_k^2)^{-1} \right.$$

$$\left.\left. - 2(2\beta + 1)z_j(1 - z_j^2) \right) \frac{\partial}{\partial z_j} \right] + p(\beta)\sum_{j=1}^{N} z_j^2 \right\} \tag{2.39b}$$

$$E(\beta) = N[\beta + \alpha(N - 1)(2\beta + 1) - \frac{1}{2}\lambda(\lambda + 1)$$

$$+ \frac{1}{3}\alpha^2(N - 1)(4N - 5)],$$

$$p(\beta) = 2\beta(2\beta + 1) - \lambda(\lambda + 1) + 4\alpha^2(N - 1)^2$$

$$+ 2\alpha(N - 1)(4\beta + 1). \tag{2.39c}$$

Singularities in Equation (2.39b) on the hyperplanes $z_j \pm z_k = 0$ cancel out if φ is symmetric with respect to any transposition of the arguments $\{z_j \leftrightarrow z_k, z_j \leftrightarrow -z_k\}$. Two structures are possible for φ, satisfying this requirement:

$$\varphi^{(1)} = \varphi_1(z_1^2, \ldots, z_N^2),$$

$$\varphi^{(2)} = \left(\prod_{j=1}^{N} z_j \right) \varphi_2(z_1^2, \ldots, z_N^2), \tag{2.40}$$

where φ_1 and φ_2 are symmetric functions of their arguments. If the condition of symmetry is fulfilled, the solutions to Equation (2.39) corresponding to the discrete spectrum may be sought in the form of series in the variables $z_1^2, \ldots, z_N^2$. By analogy with the problem of motion of one particle in the potential $-\frac{1}{2}(\cosh x)^{-2}$ one should expect that these series will, generally, diverge on the hyperplanes $z_j^2 = 1, j = 1, \ldots, N$. The series are obviously convergent in the case when they are truncated and reduce to polynomials, which may occur only at certain values of the energy and parameter β. From the normalization condition of the wave functions (2.37)

$$\int_{-1}^{1} \prod_{j=1}^{N} dz_j (1 - z_j^2)^{2\beta-1} \prod_{j>k}^{N} |Z_j - z_k^2|^{2\alpha} |\varphi(z_1, \ldots, z_N)|^2 < \infty,$$

it follows that the admissible values of parameter β are limited by condition

$$\beta > 0. \tag{2.41}$$

(We assume that at least on one of the hyperplanes $z_j^2 = 1$ φ does not vanish.) Inserting Equation (2.40) into (2.39a) and changing the variables $z_j^2 = 1 - t_j, j = 1, \ldots, N$ we arrive at the following equations for the functions φ_1 and φ_2:

$$\left\{ \sum_{j=1}^{N} \left[2\alpha t_j^2(t_j - 1) \frac{\partial^2}{\partial t_j^2} + \left(4\alpha t_j^2(t_j - 1) \sum_{k \neq j}^{N} (t_j - t_k)^{-1} \right. \right. \right.$$

$$\left. \left. \left. - 2(2\beta + 1)t_j + (4\beta + 3)t_j^2 \right) \frac{\partial}{\partial t_j} + \tilde{p}_1(\beta)t_j \right] \right\} \varphi_1$$

$$= [E - E(\beta))]\varphi_1, \tag{2.42a}$$

$$\left\{ \sum_{j=1}^{N} \left[2t_j^2(t_j - 1) - \frac{\partial^2}{\partial t_j^2} + \left(4\alpha t_j^2(t_j - 1) \sum_{k \neq j}^{N} (t_j - t_k)^{-1} \right. \right. \right.$$

$$\left. \left. \left. - 2(2\beta + 1)t_j + (4\beta + 5)t_j^2 \right) \frac{\partial}{\partial t_j} + \tilde{p}_2(\beta)t_j \right] \right\} \varphi_2$$

$$= [E - E(\beta)]\varphi_2, \tag{2.42b}$$

where

$$\tilde{p}_1(\beta) = \beta(2\beta+1) - \frac{1}{2}\lambda(\lambda+1) + \alpha(4\beta+1)(N-1) + 2\alpha^2(N-1)^2,$$

$$\tilde{p}_2(\beta) = (\beta+1)(2\beta+1) - \frac{1}{2}\lambda(\lambda+1) + \alpha(4\beta+3)(N-1)$$
$$+ 2\alpha^2(N-1)^2,$$

$$E(\beta) = -\frac{1}{2}N\left\{\frac{1}{3}\alpha^2(N^2-1) + [2\beta + \alpha(N-1)]^2\right\}. \tag{2.43}$$

We will look for solutions to Equations (2.42a) and (2.42b) in the form of expansions in powers of polynomials $\{a_j\}, j = 1, \ldots, N,$

$$a_1 = \prod_{j=1}^{N} t_j, \quad a_l = \frac{1}{(l-1)!}\hat{D}^{l-1}a_1, \quad \hat{D} = \sum_{j=1}^{N}\frac{\partial}{\partial t_j}, \quad j = 1, \ldots, N.$$
$$\tag{2.44}$$

In view of the manifest symmetry of the new variables $\{a_j\}$, in Equation (2.44), singularities of the type $(t_j - t_k)^{-1}$ disappear as it was shown in the previous section. Direct calculation with the use of the explicit form of a_j (Equation (2.44)) shows that the cancellation of singularities really takes place, and Equations (2.42a) and (2.42b) can be represented as

$$(\hat{H}_1 + \hat{G}_1)\varphi_1(a_1, \ldots, a_N) = [E - E(\beta)\varphi_1(a_1, \ldots, a_N), \tag{2.45a}$$

$$(\hat{H}_2 + \hat{G}_2)\varphi_2(a_1, \ldots, a_N) = [E - E(\beta)\varphi_2(a_1, \ldots, a_N), \tag{2.45b}$$

where

$$\hat{H}_i = 2\sum_{l,m=1}^{N}\left(\sum_{\tau=1}^{l-1}\sum_{\nu=0}^{l}(l+m-2\tau-\nu)a_\tau a_{l+m-\tau-\nu}\right)$$

$$-(N-m+2)a_{m-1}a_l - (N-l+2)a_m a_{l-1}$$

$$-(N-m+1)a_l a_m\Bigg)\frac{\partial^2}{\partial a_l \partial a_m} - \sum_{l=1}^{N}$$

$$[a_{l-1}(N-l+2)(4\beta+2l+1+2\alpha(N+l-3))$$

$$+2a_l(N-l+1)(2\beta+1+\alpha(N+l-2))]\frac{\partial}{\partial a_l} \tag{2.46}$$

$$\hat{G}_i = a_N \left(2 \sum_{l,m=1}^{N} a_l a_m \frac{\partial^2}{\partial a_l \partial a_m} \right.$$

$$\left. + \sum_{j=1}^{N} [4\alpha(N-1) + 4\beta + 2i + 1] \frac{\partial}{\partial a_l} + \tilde{p}_j \beta \right). \quad (2.47)$$

Note that in formula (2.46) we should put $a_j = 0$ for $j = 0, N+2, \ldots, a_{N+1} = 1$. Consider the conditions under which solutions to Equations (2.45a) and (2.45b) can be represented by finite-order polynomials in the variables $[a_j]$, as follows:

$$\varphi_j = \sum_{s=0}^{n} \sum_{j_1+,\ldots,j_N=s, 0 \geq j_1,\ldots,j_N \geq s} c_{j_1,\ldots,j_N}^{(i)} a_1^{j_1}, \ldots, a_N^{j_N}, \quad i = 1, 2$$

$$(2.48)$$

It is readily seen that when the operators $\hat{H}_j$ act on the polynomial $a_1^{j_1}, \ldots, a_N^{j_N}$, the degree $= j_1 + \cdots + j_N$ does not increase; the operators $\hat{G}_i$ raise that degree by one owing to the factor a_N in front of the bracket in (2.47). Therefore, the representation (2.48) may be valid only in the case when the operators $\hat{G}_i$ acting on terms in (2.47) with maximum degree n turn them into zero. From (2.47) and (2.48), it follows that this condition at given n gives rise to equations quadratic in β:

$$\tilde{P}_j(\beta) + n[4\alpha(N-1) + 4\beta + 2i + 1] + 2n(n-1) = 0. \quad (2.49)$$

From the solution to (2.49),

$$\beta_{1,2}^{(1)} = -\left[n + \alpha(N-1) + \frac{1}{4} \right] \pm \left(1 + \frac{1}{4}(1 + 2\lambda) \right),$$

$$\beta_{1,2}^{(2)} = -\left[n + \alpha(N-1) + \frac{3}{4} \right] \pm \frac{1}{4}(1 + 2\lambda), \quad (2.50)$$

one should take only those for which the condition (2.42) is valid. The parameters $\beta^{(i)}$ are defined uniquely: in formulae (2.50) the upper sign is to be taken for the second term. From (2.42), it also follows that solutions to Equations (2.45a) and (2.45b) of the type (2.48) may exist only for sufficiently large values of λ. At given λ and α, the maximal degree of the polynomials (2.48) is limited as follows:

$$\frac{1}{2}\lambda - \alpha(N-1) > n, \quad i = 1; \quad \frac{1}{2}\lambda - \alpha(N-1) > n + \frac{1}{2}. \quad (2.51)$$

When $n = 0$, the solution $\varphi_1 = \text{const}$, according to (2.38) and (2.39), corresponds to the ground-state wave function of the considered system with the Hamiltonian (2.5), where $\beta^{(1)} = \frac{1}{2}\lambda - \alpha(N-1)$. The solution $\varphi_2 = \text{const}$ defines, according to (2.4), the simplest wave function of an excited state with energy $E(\beta(2) = -\frac{1}{2}N\{\alpha^2(N^2 - 1) + [\lambda - \alpha(N-1) - 1]^2\}$. When $n > 0$, the substitution of (2.48) into (2.45a) and (2.45b) leads to systems of $\frac{(N+n)!}{N!n!}$ homogeneous algebraic equations for the coefficients $c_{j_1,\ldots,j_N}^{(i)}$ if the conditions (2.51) are satisfied.

The system possesses non-trivial solutions only for certain values of the parameter E forming a discrete spectrum. The values have to be determined from the condition for the determinants of systems to be zero. Then the problem of determining the wave functions of excited states of the discrete spectrum of an N-dimensional Schrödinger equation with the Hamiltonian (2.37) reduces to an algebraical problem; there exist two sets of wave functions corresponding to solutions of (2.45a) and (2.45b) of the type (2.48) with the constraints (2.51) on n.

In conclusion, we note that at $n = 1$ the above wave functions can be calculated explicitly (up to a normalization factor). Also, we report the corresponding energy levels:

$$\varphi_{1,l}^{(1)} = c_{1,l}^{(1)} \sum_{j=1}^{l} (-1)^{l-j} \frac{(N-j+1)!}{(N-l+1)!(l-j)!} \frac{\Gamma(2\lambda-1)/2\alpha - N + l)}{\Gamma(2\lambda-1)/2\alpha - N + j)}$$

$$\times \frac{\Gamma((\lambda-1/\alpha - 2N + l + j - 1)}{\Gamma((\lambda-1)/\alpha - 2N + 2l - l)} a_j,$$

$$E_{1,l}^{(1)} = \frac{1}{2}N\left\{[\lambda - 2 - \alpha(N-1)]^2 + \frac{1}{3}\alpha^2(N^2 - 1)\right\}$$

$$+ 2(N-l+1)[\lambda - 1 - \alpha(N-l)], \quad l = 2,\ldots,N+1$$

$$\varphi_{2,l}^{(1)} = c_{2,l}^{(1)} \sum_{j=1}^{l} (-1)^{l-j} \frac{(N-j+1)!}{(N-l+1)!(l-j)!} \frac{\Gamma(2\lambda-1)/2\alpha - N + l)}{\Gamma(\lambda - 2\alpha - 2N + 2l - 1)}$$

$$\times \frac{\Gamma((\lambda-2)/\alpha - 2N + j + l - 1)}{\Gamma((\lambda-2)/\alpha - 2N + 2l - 1)} a_j,$$

$$E_{2,l}^{(1)} = \frac{1}{2} N \left\{ [\lambda - 3 - \alpha(N-1)]^2 + \frac{1}{3}\alpha^2(N^2 - 1) \right\}$$

$$+ 2(N - l + 1)[\lambda - 2 - \alpha(N - l)], \quad l = 1, \ldots, N+1.$$

These functions can be calculated explicitly because at $n = 1$ the matrices of the operators $\hat{H}_j + \hat{G}_i$ in the basis $\{a_1, \ldots, a_N, 1\}$ have non-zero elements only on the principal and neighboring diagonals. In the general case $n \geq 2$, the structure of the matrices $\hat{H}_i + \hat{G}_i$ is more complicated. The problem of constructing the wave functions φ_l explicitly for $n \geq 2$ can no longer be reduced to simple recurrence relations for the coefficients in (2.48) and deserves further study.

2.7. The Discrete Spectrum of Quantum Systems in the Pöshl–Teller Potential

Our last example is finding the discrete spectrum of the quantum system in the Pöshl–Teller potential since nobody knows how to treat the general potential created by the particle with infinite mass:

$$g_1(\sinh x)^{-2} + g_2(\cosh x)^{-2} + g_3 \cosh(2x) + g_4 \cosh(4x).$$

We will restrict ourselves to the more simple potential of the external field:

$$W(x) = \frac{\mu(\mu - 1)}{2}(\sinh x)^{-2} - \frac{\lambda(\lambda - 1)}{2}(\cosh x)^{-2}.$$

The light particles interact via the potential

$$V(x_j - x_k) + V(x_j + x_k), \quad V(x) = \alpha(\alpha - 1)(\sinh(x))^{-2}. \quad (2.52)$$

So, the Hamiltonian under our consideration will be [50]

$$H = \sum_{1}^{N} \left(\frac{p_j^2}{2} + W(x_j) \right) + \sum_{j>k} [V(x_j - x_k) + V(x_j + x_k)]. \quad (2.53)$$

Note that a recent investigation of bound states and scattering of one particle in the Pöshl–Teller potential has testified to their tight relation with irreducible representations of $SU(2)$ and $SU(1,1)$ groups. We shall, however, concentrate on excited states of the discrete spectrum leaving the continuous spectrum for further investigation and shall show how to calculate both the energy levels (their number is finite and is determined by the parameters of the Hamiltonian (2.52) (λ, μ, α)) and the wave functions normalized by the condition

$$\int dx_1, \ldots, dx_N |\psi(x_1, \ldots, x_N)|^2 = 1. \tag{2.54a}$$

We will solve this problem by a method that makes no use of the explicit form of extra constants of motion, polynomials in momenta of a high degree. Our method proposed in previous sections will be based on the expansion of solutions to the Schrödinger equation into a series of a special form over functions symmetric in the variables $\{x_j\}$. As in the previous section, we shall suppose that particles obey the Bose statistics.

Upon these introductory remarks we proceed to consider the Schrödinger equation $H\psi = E\psi$ having made the following change of variables: $z = \tanh x_j, j = 1, \ldots, N$ (when there is no interaction with potential $V(x)$, a substitution like that transforms the equations of motion of particles into hypergeometrical forms):

$$\sum_{j=1}^{N} \left\{ -\frac{1}{2} \left[(1 - z_j^2) \frac{\partial}{\partial z_j} (1 - z_j^2) \frac{\partial}{\partial z_j} + (1 - z_j^2) \right. \right.$$

$$\left. \left. \times \left(\lambda(\lambda + 1) - \mu(\mu - 1) z_j^{-2} \right) \right] \right\}.$$

$$+ \sum_{j>k} N \frac{2\alpha(\alpha - 1)(1 - z_j^2)(1 - z_k^2)(z_j^2 + z_k^2)}{(z_j^2 - z_k^2)^2} \psi = E\psi.$$

This equation has singularities on hyperplanes $z_j = 0$ and $z_j \pm z_k = 0, j, k = 1, \ldots, N$ splitting an N-dimensional space into a set of polyhedral angles. Since potential barriers of the form $gz^{-2}, g \geq 3/4$ are absolutely impenetrable, we may consider this equation in any of those angles (on the boundary of which the wave function should vanish), extending then the results obtained to other angles in a symmetric manner. For definiteness, we fix the polyhedral angle by the condition $\{z_j > 0, z_j \pm zk > 0, j > k\}$. Now, let us lower the order of singularities in this equation. Setting

$$\psi(z_1, \ldots, z_N) = \chi(z_1, \ldots, z_N)\Phi_{\alpha\mu}^{(\beta)}(z_1, \ldots, z_N), \qquad (2.54)$$

where

$$\Phi_{\alpha\mu}^{(\beta)} = \prod_{j=1}^{N} z_j^{\mu}(1 - z_j^2)^{\beta} \prod_{j>k}(z_j^2 - z_k^2)^{\alpha}, \qquad (2.55)$$

and changing the variables $z_j^2 = q_j$, $j = 1, \ldots, N$ we get for $\chi(z_1, \ldots, z_N)$ the equation

$$-\left\{ \sum_{j=1}^{N} \left[2q_j(1 - q_j)^2 \frac{\partial^2}{\partial q_j^2} + \left(4\alpha q_j(1 - q_j)^2 \sum_{k \neq j}^{N}(q_j - q_k)^{-1} \right. \right. \right.$$

$$\left. + (1 - q_j)(2\mu + 1) - (4\beta + 2\mu + 3)q_j \right) \frac{\partial}{\partial q_j} \Bigg]$$

$$+ p(\beta) \sum_{j=1}^{N} q_j \Bigg\} \chi = (E - \varepsilon(\beta))\chi. \qquad (2.56)$$

Here,

$$p(\beta) = \beta(2\beta + 2\mu + 1) + \frac{(\mu - \lambda)(\mu + \lambda + 1)}{2}$$

$$+ 2\alpha^2(N - 1)^2 + \alpha(N - 1)(4\beta + 2\mu + 1),$$

$$\varepsilon(\beta) = N \left[\beta(2\mu + 1) + \alpha(N - 1)(2\beta + 2\mu + 1) \right.$$

$$\left. + \frac{(\mu - \lambda)(\mu + \lambda + 1)}{2} + \frac{\alpha^2}{3}(N - 1)(4N - 5) \right].$$

As will be shown in the following, the function $\Phi_{\alpha\mu}^{(\beta)}$ with a certain value of the parameter β represents the wave function of the ground state of the Hamiltonian (2.53) we are considering.

It is easy to see that when the terms $\sum_{k \neq j}(q_j - q_k)^{-1}$ are absent in (2.55), the variables may be separated, and the solution may be searched in the form of products of power series in q_j, which, generally, diverge as $q \to 1$. Convergence on the hyperplanes ensuring the normalization condition being fulfilled takes place provided the series are truncated and reduced to polynomials of a finite degree. This occurs only on certain values of the parameters E and β; the result is the spectrum of bound states of N non-interacting particles in the Pöshl–Teller potential.

To construct solutions to equation (2.56) corresponding to the discrete spectrum, we take advantage of the analogy with the above scheme applicable in the absence of interactions. We shall solve Equation (2.56) in the form of series over functions symmetric in any two coordinates $\{q_j, q_k\}$. According to previous sections, a symmetry of this type is necessary and sufficient for removing singularities from (2.56) when $qj \to qk$. First, it is convenient to make the change of variables $q_j \to 1 - t_j, j = 1, \ldots, N$ in Equation (2.56), reducing thus the latter to the form

$$\hat{H}(\beta)\chi = (E - \hat{\varepsilon}(\beta))\chi, \qquad (2.57)$$

where

$$\hat{H}(\beta) = \sum_{j=1}^{N} \left[2(t_j^3 - t_j^2)\frac{\partial^2}{\partial t_j^2} + \left((4\beta + 2\mu + 3)tj^2 - 2(2\beta + 1)t_j \right. \right.$$

$$\left. \left. + 4\alpha t_j^2(t_j - 1)\sum_{k \neq j}^{N}(t_j - t_k)^{-1} \right) \right] \frac{\partial}{\partial t_j} + p(\beta)\sum_{j=1}^{N} t_j,$$

$$\varepsilon(\beta) = \epsilon(\beta) - Np(\beta) = -\frac{N}{2}\left[(2\beta + \alpha(N-1))^2 + \frac{\alpha^2(N^2 - 1)}{3} \right]$$

$$(2.58)$$

Then, we substitute $\{t_j\}$ by new variables explicitly symmetric with respect to the transpositions $t_j \leftrightarrow t_k$

$$a_1 = \prod_{j=1}^{N} t_j, \quad a_l = \frac{\hat{D}^{l-1}}{(l-1)!}, \quad \hat{D} = \sum_{j=1}^{N}\frac{\partial}{\partial t_j}, \quad j = 1, \ldots, N. \quad (2.59)$$

If in what follows the index m of quantities $\{a_m\}$ will assume values less than 1 or larger than N, one should set

$$a_{N+1} = 1, \quad a_0 = a_{N+2} = a_{N+3} = \cdots 0.$$

To pass to variables $\{a_j\}$ (2.59) in (2.57) and (2.58), one should represent in the form of functions of $\{a_j\}$ the following quantities:

$$\hat{B}_l a_l, \quad B_l = \sum_{j=1}^{N} t_j^i \frac{\partial}{\partial t_j}, \quad i = 1, 2, \tag{2.60a}$$

$$\hat{C}_i a_l, \quad \sum_{j \neq k} t_j^i (t_j - t_k)^{-1} \frac{\partial}{\partial t_j}, \quad i = 2, 3, \tag{2.60b}$$

$$F_{lm}^{(i)} = \sum_{j=1}^{N} t_j^i \frac{\partial a_l}{\partial t_j} \frac{\partial a_m}{\partial t_j}, \quad i = 2, 3; \quad l, m = 1, \ldots, N. \tag{2.60c}$$

Let us first calculate (2.60a). Noting that

$$[\hat{D}, \hat{B}_l] = i\hat{B}_{l-1}, \quad \hat{B}_0 = \hat{D},$$

and using the definition (2.59) of a_l, we find

$$((l-1)!)^{-1} \hat{B}_i \hat{D}^{l-1} a_1 = ((l-1)!)^{-1}$$

$$\times \left[\sum_{j=0}^{l-2} \hat{D}^j [\hat{B}_i \hat{D}] \hat{D}^{l-2-j} a_1 + \hat{D}^{l-1} \hat{B}_i a_1 \right]. \tag{2.61}$$

Since

$$\hat{B}_1 = N a_1, \quad \hat{B}_2 a_1 = a_N a_1,$$

$$\hat{D}^{l-1} a_N a_1 = a_N \hat{D}^{l-1} a_1 + N(l-1) \hat{D}^{l-2} a_1,$$

we get from the definition of $\hat{B}$ and $\hat{D}$:

$$\hat{B}_1 a_l = (N - l + 1) a_l, \quad \hat{B}_2 a_l = a_N a_l - (N - l + 2) a_{l-1}.$$

Now we shall calculate (2.60b). First, consider the case $l = 1$, then

$$\hat{C}_i a_1 = a_1 \sum_{j \neq k}^{N} (t_j - t_k)^{-1} t_j^{l-1}.$$

Calculating the total sum (2.61), we arrive at the relations

$$\hat{C}_1 a_1 = 0, \quad \hat{C}_2 a_1 = \frac{N(N-1)}{2}, \quad \hat{C}_3 a_1 = a_N a_1 (N-1).$$

Note that the commutators of operators $\hat{C}_i$ and $\hat{D}$ have also the simple form

$$[\hat{C}_i, \hat{D}] = -i\hat{C}_{i-1}, \quad [\hat{C}_0, \hat{D}] = 0. \tag{2.62}$$

Successively applying (2.62) in identities analogous to (2.61),

$$\hat{C}_i \hat{D}^{l-1} a_1 = \sum_{j=0}^{l-2} \hat{D}^j [\hat{C}_j, \hat{D}] \hat{D}^{l-2-j} a_1 + \hat{D}^{l-1} \hat{C}_i a_1,$$

or using the scheme of the proof described in detail in Section 2.5, we obtain

$$\hat{C}_2 a_l = \frac{1}{2}(N-l+1)(N+l-2)a_l,$$

$$\hat{C}_3 a_l = (N-1)a_N a_l - \frac{1}{2}(N-l+2)(N+l-3)a_{l-1}. \tag{2.63}$$

And finally, let us derive explicit expressions for the quantities $F_{lm}^{(i)}$ (2.60c). Since the operator $[\hat{D}, t_j(\partial/\partial t_j)] = \partial/\partial t_j$ commutes with $\hat{D}$, the definition of polynomials $\{a_l\}$ (2.61) allows us to find their property

$$t_j \frac{\partial a_l}{\partial t_j} = a_l - \frac{\partial a_{l-1}}{\partial t_j}, \tag{2.64}$$

which makes it possible to express $F_{lm}^{(3)}$ in terms of $F_{lm}^{(2)}$. Indeed, grouping factors in (2.60c) and applying (2.64), we get

$$F_{lm}^{(3)} = \sum_{j=1}^{N} \left(t_j \frac{\partial a_l}{\partial t_j} \right) t_j^2 \frac{\partial a_m}{\partial t_j} = \sum_{j=1}^{N} \left(a_l - \frac{\partial a_{l-1}}{\partial t_j} \right) t_j^2 \frac{\partial a_m}{\partial t_j}$$

$$= a_l \hat{B}_2 a_m - F_{l-1,m}^{(2)} = a_N a_l a_m$$

$$-(N-l+2)a_{m-1}a_l - (N-m+2)a_{l-1}a_m. \tag{2.65}$$

Since, the left-hand side of (2.65) is symmetric with respect to the transposition of indices l and m, the same property should also hold for the right-hand side. This leads to the system of equations for the quantities $F_{lm}^{(2)}$:

$$F_{l-1,m}^{(2)} - F_{l,m-1}^{(2)} = (N-l+2)a_{m-1}a_l - (N-m+2)a_{l-1}a_m. \quad (2.66)$$

It is also not difficult to derive an explicit expression for $F_{1,l}^{(2)}$. Since $\frac{\partial a_1}{\partial t_j} = t_j^{-1}a_1$, then

$$F_{1,l}^{(2)} = \sum_{j=1}^{l} t_j^2 \frac{\partial a_1}{\partial t_j} \frac{\partial a_l}{\partial t_j} = a_1 \hat{B}_1 a_l = (N-l+1)a_1 a_l. \quad (2.67)$$

Using (2.67) for a successive lowering of the first index of $F^{(2)}$ and expression (2.66), we can derive $F_{lm}^{(2)}$ explicitly as follows:

$$F_{lm}^{(2)} = (N-m+1)a_l a_m - \sum \tau = 1^{l-1}(l+m-2\tau)a_\tau a_{l+m-\tau}. \quad (2.68a)$$

From (2.60a) and (2.66), we get the expression for $F_{lm}^{(3)}$ as follows:

$$F_{lm}^{(3)} a_l a_m a_N - (N-m+2)a_{m-1}a_l - (N-l+2)a_{l-1}a_m$$
$$+ \sum_{\tau=1}^{l-1}(l+m-2\tau-1)a(\tau)a_{l+m-\tau-1}. \quad (2.68b)$$

Allowing for (2.60a) and (2.60b) and the relation $\partial^2 a_l/\partial t^2 = 0$, $l,j = 1,\ldots,N$, we can write Equations (2.57) and (2.58) in terms of the variables $\{a_1,\ldots,a_N\}$:

$$(\hat{H}_1(\beta) + \hat{H}_2(\beta))\chi = (E - \tilde{\varepsilon}(\beta))\chi \quad (2.69)$$

$$\hat{H}_1(\beta) = 2 \sum_{l,m=1}^{N} \left[\sum_{\tau=1}^{l-1}\sum_{v=0}^{1}(l+m-2\tau-v)a_\tau a_{l+m-\tau-v} \right.$$
$$- (N-m+2)a_{m-1}a_l - (N-l+2)a_m a_{l-1}$$
$$\left. - (N-m+1)a_l a_m \right] \frac{\partial^2}{\partial a_l \partial a_m} - \sum j = 1^N [a_{l-1}(N-l+2)$$

$$\times (4\beta + 2\mu + 3 + 2\alpha(N + l - 3))$$

$$+ 2a_l(N - l + 1)(2\beta + 1 + \alpha(N + l - 2))]\frac{\partial}{\partial a_l}, \qquad (2.70a)$$

$$\hat{H}_2(\beta) = a_N \left[2\sum_{l,m}^{N} a_l a_m \frac{\partial^2}{\partial a_l \partial a_m} \right.$$

$$\left. + \sum_{l=1}^{N} a_l \frac{\partial}{\partial a_l}(4\beta + 2\mu + 3 + 4\alpha(N - 1)) + p(\beta) \right]. \qquad (2.70b)$$

It is just this form of the Schrödinger equation that is most suitable for searching solutions obeying the normalization condition (2.55a), which for the functions χ, according to (2.57), (2.58) has the form

$$\int_{-1}^{1} \prod_{j=1}^{N} z_j^{\alpha}(z_j)^{2\mu}(1 - z_j^2)^{2\beta-1} \prod_{j>k}^{N} |z_j^2 - z_k^2|^{2\alpha}|\chi|^2 = 1. \qquad (2.71)$$

It is natural to look for solutions to Equation (2.69) in the form of power series in $\{a_j\}$,

$$\chi = \sum_{s=0}^{n} \sum_{0 \leq j_1,\ldots,j_N \leq s, j_1+,\ldots,+j_N=s} c_{j_1,\ldots,j_N}^{(s)} a_1^{j_1},\ldots,a_N^{j_N} \qquad (2.72)$$

From the analysis of the structure of operators $\hat{H}_{1,2}(\beta)$, it is clear that in the general case series (2.70a) and (2.70b) include an infinite number of terms and may diverge in some regions of the space $\{a_1,\ldots,a_N\}$. The operator $\hat{H}_2(\beta)$ acting on the polynomial $a_1^{j_1},\ldots,a_N^{j_N}$ of degree s transforms it into a polynomial of degree $(s + 1)$ so that the chain of equations for $c_{j_1,\ldots,j_N}^{(s)}$ arising upon substituting (2.72) into (2.70) is, in the general case, infinite. The truncation of summation over s and reduction of series (2.72) to polynomial without singularities may occur only in the case when the operator $\hat{H}_2(\beta)$ acting on the polynomial of maximum degree n in (2.72) reduces it to zero:

$$\hat{H}_2(\beta)a_1^{j_1} \cdots a_N^{j_N} = 0, \quad j_1 + \cdots + j_N = n. \qquad (2.73)$$

From (2.70b) and (2.56), it follows that Equation (2.73) is equivalent to the following condition to be imposed on the parameter β:

$$2\beta^2 + \beta(2\mu + 1 + 4n + 4\alpha(N - 1)) + n(2\mu + 2n + 1)$$
$$+ \alpha(N - 1)(2\mu + 4n + 1 + 2\alpha)(N - 1))$$
$$+ \frac{1}{2}(\mu - \lambda)(\mu + \lambda + 1) = 0. \tag{2.74}$$

This condition, together with the normalization condition (2.71), has the solution

$$\beta(n) = \frac{\lambda - \mu}{2} - \alpha(N - 1) - n. \tag{2.75}$$

From (2.71) it also follows that $\beta > 0$, i.e. determining the maximum degree of polynomial (2.72) (and, consequently, the number of levels of the discrete spectrum is, according to (2.75), limited:

$$n < \frac{\lambda - \mu}{2} - \alpha(N - 1). \tag{2.76}$$

We denote the maximal number satisfying inequality (2.76) by $\tilde{n}$. The set of all polynomials of the variables $\{a_j\}$ of a degree not higher than $\tilde{n}$ composes a linear space L of the dimensionality $(N + \tilde{n})!/N!\tilde{n}!$. In this space, $\hat{H}_1(\beta(\tilde{n}))$ and $\hat{H}_2(\beta(\tilde{n}))$ are linear operators. The total number of levels of the discrete spectrum of Hamiltonian (2.52) equals the dimensionality of the above linear space L, $(N + \tilde{n})/N!\tilde{n}!$; the level themselves are defined by eigenvalues of the matrix $H_1(\beta(\tilde{n}))$ and a constant

$$\varepsilon(\beta\tilde{n})) = -\frac{N}{2}\left[\frac{\alpha^2}{3}(N^2 - 1) + (\lambda - \mu - \alpha(N - 1) - 2\tilde{n})^2\right].$$

Consider the structure of this matrix in the basis of the space L formed by elements $\{a_1^{j^1} \cdots a_N j^N\}$

$$0 \leq j_1, \ldots, j_N \leq \tilde{n}, \quad j_1 + \cdots + j_N \leq \tilde{n}.$$

The ordering in this basis is introduced by the following rule. First, we split the whole set of basis elements into subsets L_k defined by the conditions $j_1 + \cdots + j_N = k, k = 0, 1, \ldots, \tilde{n}$ and assume the elements of L_k have a smaller number than elements of $L_{k'}$, if $k > k'$. Then the matrix $\hat{H}_1(\beta(\tilde{n})) + \hat{H}_2(\beta(\tilde{n}))$ splits into $(\tilde{n}+1)^2$ blocks, which will be denoted by $\hat{H}_{kk'}$ (in what follows, that type of representation of the matrix will be called the block representation). Second, we establish the ordering in each of the subsets L_k containing $(N+k-1)!/(N-1)!k!$ elements: if for two elements from $L_k, a_1^{j_1}, \ldots, a_N^{j_N}, a_1^{j'_1}, \ldots, a_N^{j'_N}$ there hold the relations $j_1 = j'_1, \ldots, j_s = j'_s, j_{s+1} > j'_{s+1}$, then the first element has a smaller number.

Note that with the ordering thus defined, the matrix of operator $\hat{H}_2(\beta(\tilde{n}))$ is an upper triangular matrix, and in the block representation non-zero are only the matrix elements $(\hat{H}_2)_{k+1,k}$ as well as $(\hat{H}_1)_{kk}$ and $(H_1)_{k-1,k}$. The latter, breaking the triangular nature of the whole matrix $(\hat{H}_1 + \hat{H}_2)$ in the block representation, results from the structure of the first term in $\hat{H}_1(\beta)$ (2.70a) when $l+m \geq N+2$ in the sum over τ, ν, the index of factor $a_{l+m-\tau-\nu}$ can assume the value $N+1$. Since $a_{N+1} = 1, \hat{H}_1(\beta)$ contains terms of the form $(2(N+1) - l - m)a_{l+m-N-1}(\partial^2)/(\partial a_l \partial a_m)$ lowering the degree of polynomial $a_1^{j_1}, \ldots, a_N^{j_N}$.

It is not difficult to prove that all the matrices $(\hat{H}_1)_{kk}$ are upper triangular: the action of operators $a_{l-1}\partial/\partial a_l, a_{m-1}al(\partial^2/(\partial a_l \partial a_m),$

$$\sum_{l>m}^{N} \sum_{\tau=1}^{m-1} \sum_{\nu=0}^{1}(l + m - 2\tau - \nu)a_\tau a_{l+m-\tau-\nu}(\partial^2)/(\partial a_l \partial a_m),$$

on the element $a_1^{j_1} \cdots a_N^{j_N}$ produces elements with lower numbers in the same subset L_k.

Let us show that despite the non-zero elements $(\hat{H}_1)_{k-1,k}$, the whole matrix $\hat{H}_1(\beta(\tilde{n})) + \hat{H}_2(\beta(\tilde{n}))$ can be made triangular. Note first that the block $(\hat{H}_1)_{01}$ consists only of zeros, and the matrix $(\hat{H}_1)_{11}$ can be transformed into a diagonal form by subtracting columns, the procedure does not change its diagonal elements. The block $(\hat{H}_1)_{12}$ already contains non-zero elements in some columns (and rows $\{l\}$). They can be made zero by subtracting from these columns of the whole matrix $\hat{H}_1 + \hat{H}_2$ those which passing through the block $(\hat{H}_1)_{12}$ have there diagonal elements corresponding to rows $\{l\}$. And owing to elements of the block $(\hat{H}_2)_{12}$, new terms appear in the matrix $(\hat{H}_1)_{22}$. However, because of the way of ordering we have chosen, these terms do not change the structure of the matrix $(H_1)_{22}$: it remains, as before, upper triangular, and its diagonal elements are also unchanged. Continuing this procedure one may successively eliminate all non-zero elements from the blocks $(\hat{H}_1)_{k-1,k}$ without changing the diagonal elements of matrices $(\hat{H}_1)_{kk}$ which are eigenvalues of the matrix $(\hat{H}_1(\beta(\tilde{n})) + \hat{H}_2(\beta(\tilde{n})))$. Direct calculation by formula (2.70a) thus shows that the discrete spectrum of Hamiltonian (2.52) is defined by the set of N integers $\{j_1, \dots, j_N\}, 0 \geq j_1, \dots, j_N \geq \tilde{n}, j_1 + \cdots + j_N = s \geq \tilde{n}$; the energy levels, up to $\varepsilon(\beta(\tilde{n}))$, are diagonal elements of the operator

$$-2\left\{\sum_{l>m}(N-l+1)a_l a_m(\partial^2)/(\partial a_l \partial a_m) + \sum_l (N-l+1)a_l^2(\partial^2)/(\partial a_l^2\right.$$

$$\left. -2\sum_l a_l(N-l+1)(2\beta(\tilde{n})+1+\alpha(N+l-2))(\partial)/(\partial a_l\right\},$$

in the basis $\{a_1^{j_1} \cdots a_N^{j_N}\}$:

$$-E_{j_1 \cdots j_N} = \frac{N}{2}\left[\frac{\alpha^2}{3}(N^2-1) + (\lambda - \mu - \alpha(N-1) - 2\tilde{n})^2\right]$$

$$+2\left[s(N+1)(s+\lambda-\mu-\alpha(N-1)-2\tilde{n}-\alpha N),\right.$$

$$+\sum_{l=1}^{N} l j_l(\alpha(2N+1)+2\tilde{n}+\mu-\lambda-j_l+\alpha l)$$

$$\left. +2\sum_{l>m} l j_l j_m\right]. \tag{2.77}$$

Note that the parameters of Hamiltonian (2.52) enter into (2.77) only in the combination $(\lambda - \mu)$. Therefore, one can immediately define the discrete spectrum of the Sutherland hyperbolic system in the Morse oscillator in the limit

$$\lambda \to \frac{A}{2}\exp(2q) + A - 1/2, \quad \mu \to \frac{A}{2}\exp(2q) - A + \frac{1}{2},$$

$$x_j \to x_j - q(j=1,\dots,N), \quad q \to +\infty.$$

Substitution of this limit to (2.77) gives the required result. The ground-state energy may be calculated at $j_1 = \tilde{n}, j_2 = \cdots j_N = 0$ by the formula (2.77):

$$E^{(0)} = -\frac{N}{2}\left[\frac{\alpha^2}{3}(N^2 - 1) + (\lambda - \mu - \alpha(N - 1))^2\right].$$

Once the levels of discrete spectrum (2.77) are found, the problem of finding wave functions reduces to the substitution of (2.77) and (2.72) into (2.69) and the solution of the systems of $(N + \tilde{n})!/N!\tilde{n}!$ homogeneous linear algebraic equations with zeroth determinant. For $\tilde{n}$ arbitrary, we could not find these solutions in an explicit form. An exception is the case $\tilde{n} = 1$ when the functions χ are of the form

$$\chi = \sum_{j=1} C_j a_j + C_{N+1}$$

and matrix elements of terms in $\hat{H}_1(\beta(1)$ containing double differentiation with respect to variables $\{a_j\}$ vanish. The system of equations for coefficients C_j can be written in the following form:

$$(E - \varepsilon(\beta(1)))C_{N+1} = 0,$$

$$(N - j + 1)[2\lambda - 1 - 2\alpha(N - j)]C_{j+1}$$
$$+ [2(N - j + 1)(\lambda - \mu - 1 - \alpha(N - j))$$
$$+ E - \varepsilon(\beta(1))]C_j = 0.$$

Replacing here E by energy levels at $n = 1$,

$$E(2) = \varepsilon(\beta(1)) - 2(N - l + 1)(\lambda - \mu - 1 - \alpha(N - j),$$

we get

$$(N - l + 1)C_{N+1}^{(l)} = 0,$$

$$(N - j + 1)[2\lambda - 1 - 2\alpha(N - j)]C_{j+1}^{(l)}$$
$$-2(j - l[\alpha(j + l - 2N - 1) + \lambda - \mu - 1]C^{(l)} = 0,$$

$$l = 2, \ldots, N + 1 \qquad (2.78)$$

Direct calculation of the coefficients $C_j^{(l)}$ by the formula (2.78) allows the wave functions $\chi(l)$ at $n = 1$ to be found explicitly (up to a normalization constant)

$$\chi_l^{(1)} = d_l \sum_{j=1}^{l} (-1)^{l-j} \frac{(N-j+1)! \Gamma\left(\frac{2\lambda-1}{2\alpha} - N + l\right)}{(N-l+1! (l-j)! \Gamma\left(\frac{2\lambda-1}{2\alpha} - N + j\right)}$$

$$\times \Gamma\left(\frac{\lambda-\mu-1}{\alpha} - 2N + l + j - 1\right)$$

$$\times \Gamma\left(\frac{\lambda-\mu-1}{\alpha} - 2N + 2l - 1\right) \times a_j. \tag{2.79}$$

The ground-state wave function of a system with Hamiltonian (2.52) is of the form (2.55)

$$\Phi_{\alpha\mu}^{((\lambda-\mu)-\alpha(N-1))} = Z^{-1/2} \prod_{j=1} z_j^\mu (1 - z_j^2))^{1/2(\lambda-\mu)-\alpha(N-1)}$$

$$\times \prod_{j>k}^{N} (z_j^2 - z_k^2)^\alpha. \tag{2.80}$$

The integral defining the normalizing factor Z in (2.80) was calculated by Dotsenko and Fateev [41] as follows:

$$Z = \prod_{j=1}^{N} \frac{\Gamma(j\alpha)}{\Gamma(\alpha)} \prod_{j=0}^{N} \frac{\Gamma(\mu + \frac{1}{2} + j\alpha)}{\Gamma(\lambda + \frac{1}{2} + j\alpha)} \Gamma(\lambda - \mu - \alpha(N - j - 1)).$$

For functions (2.80), we could not derive explicit expressions for d_l.

Note that for $0 < (\lambda - \mu/2) - \alpha(N - 1) < 2$, the quantities $\chi_l^1 \Phi_{\alpha\mu}^{((\lambda-\mu/2)-\alpha(N-1)-1)}$ and $\Phi_{\alpha\mu}^{((\lambda-\mu/2)-\alpha(N-1))}$ (2.80) form a complete set of wave functions of the discrete spectrum of our Hamiltonian. In the case of physical interest (systems in the Morse potential), the wave functions of the discrete spectrum at $n = 1$ can be found by solving the system of equations analogous to (2.79) as follows:

$$\psi_1(l)(x_1, \ldots, x_N) = \tilde{d}_l \prod_{s=1}^{N} \exp(x_s)(2A - 2\alpha(N-1) - 3)$$

$$- Al^{2x_s} \prod_{j>k}^{N} |e^{2x_j} - e^{2xk}|^{\alpha} \times \sum_{j=1}^{l} \frac{(-1)^{l-j}}{(j-1)!} \left(\frac{2A}{\alpha}\right)^{l-j}$$

$$\times \frac{(N-j+1)!}{(l-j)!(N-l+1)!}$$

$$\times \frac{\Gamma((2A-2)/\alpha - 2N + j + l - 1)}{\Gamma((2A-2)/\alpha - 2(N-l) - 1)}.$$

Thus, we have explicitly found the energy levels and the simplest wave functions of excited states of the discrete spectrum of the original Hamiltonian (2.52) and the Hamiltonian of the interacting light particle system in the Morse potential created by one particle of infinite mass. We have not utilized the explicit form of extra constants of motion of these Hamiltonians. Therefore, it is natural that the method I have applied cannot be extended to more complicated Hamiltonians introduced in Chapter 1 on the classical level, which turn to be also quantum integrable. In the case (2.52), the ground-state wave function can be represented in a factorized form like (2.72) only at certain constraints on parameters of the potential of an external field; no information is available on the continuum part of the spectrum of these Hamiltonians. Probably, it will be possible to include the systems with the Hamiltonians found in Chapter 1 into the scheme of the quantum inverse scattering problem the universality of which has been demonstrated by a successful reduction of the problem of the spectrum of the quantum periodic N-particle Toda lattice. A program like that being realized even in the simplest non-trivial case $N = 2$ would provide, as we see it, an essential progress in studying the quantum systems with the potential of an external field and in understanding the nature of the symmetry of the Schrödinger equation that leads to a comparatively simple expression (2.70) for the discrete spectrum levels of integrable systems with the Morse potential.

Chapter 3

Integrable Systems of Quasi-particles

3.1. Introduction

The idea of spin exchange interaction of electrons as a natural explanation of ferromagnetism was first proposed by Heisenberg [42] and soon realized in mathematical form by Dirac [43]. But the first appearance of the famous Heisenberg Hamiltonian in solid-state physics occurred three years later in the book by van Vleck [44]. Now it is of common use and was investigated from many points of view by various methods of condensed matter theory. In two and higher dimensions, the problem of finding the eigenvalues and eigenvectors can be solved only by approximate or numerical methods. In one dimension, the *exact* solution was obtained in the seminal paper by Bethe [45] who considered the most important case of nearest-neighbor exchange described by the Hamiltonian

$$H = \sum_{1 \leq j \neq k \leq N} h(j - k)(\vec{\sigma}_j \vec{\sigma}_k - 1), \tag{3.1}$$

where $\{\sigma_j\}$ are the usual Pauli matrices acting on the $s = 1/2$ spin located at the site j and *exchange constants* $\{h\}$ are of extreme short-range form, as follows:

$$h(j) = J(\delta_{|j|,1} + \delta_{|j|,N-1}). \tag{3.2}$$

It turned out that the solution comes in the form of linear combinations of plane waves chosen as to satisfy certain conditions required by (3.1) and (3.2).

Starting with this solution known as *Bethe Ansatz*, the investigation of one-dimensional exactly solvable models of interacting objects (spins, classical and quantum particles) has given a number of results both of physical and mathematical significance [46]. Bethe found his solution empirically; at that time the possibility of solving the quantum-mechanical problems was not associated with the existence of underlying symmetry. The role of such symmetries was recognized much later, with one of the highlights being the Yang-Baxter equation which allows one to find some regular ways to find new examples of exactly solvable models [47, 48]. In many cases, however, the empirical ways are more productive since they use some physical information on their background and are not so complicated from the mathematical viewpoint.

This concerns especially the Calogero–Sutherland–Moser (CSM) models which were discovered about 40 years ago. They describe the motion of an arbitrary number of classical and quantum nonrelativistic particles interacting via two-body singular potentials with the Hamiltonian

$$H_{CSM} = \sum_{j=1}^{M} \frac{p_j^2}{2} + l(l+1) \sum_{j<k}^{M} V(x_j - x_k), \qquad (3.3)$$

where $\{p, q\}$ are canonically conjugated momenta and positions of particles, $l \in \mathbf{R}$ and the two-body potentials are of the following form:

$$V(x) = \frac{1}{x^2}, \quad \frac{k^2}{\sin^2(\kappa x)}, \qquad (3.4)$$

$$V(x) = \frac{\kappa^2}{\sinh^2(\kappa x)}, \qquad (3.5)$$

$$V(x) = \wp(x), \qquad (3.6)$$

where $\kappa \in \mathbf{R}_+$ and $\wp(x)$ is the double periodic Weierstrass $\wp$ function determined by its two periods $\omega_1 \in \mathbf{R}_+$, $\omega_2 = i\pi/\kappa$ as

$$\wp(x) = \frac{1}{x^2} + \sum_{m,n\in\mathbf{Z}, m^2+n^2\neq 0} \left[\frac{1}{(x - m\omega_1 - n\omega_2)^2} - \frac{1}{(m\omega_1 + n\omega_2)^2} \right].$$
$$(3.7)$$

The solvability of the eigenproblem for first two potentials (3.4) has been found independently by Calogero [49] and Sutherland [50] in the quantum case while (3.5) and (3.6) were found much later [51,52] for classical particles by constructing extra integrals of motion (conserved quantities) via the method of Lax pair. Therefore, it turned out that the dynamical equations of motion are equivalent to the $(M \times M)$ matrix relation

$$\frac{dL}{dt} = [L, M], \tag{3.8}$$

where

$$L_{jk} = p_j \delta_{jk} + (1 - \delta_{jk}) f(x_j - x_k),$$

$$M_{jk} = (1 - \delta_{jk}) g(x_j - x_k) - \delta_{jk} \sum_{m \neq j}^{M} V(x_j - x_m), \tag{3.9}$$

if the functions f, g, V obey the Calogero–Moser functional equation

$$f(x)g(y) - f(y)g(x) = f(x + y)[V(y) - V(x)], \tag{3.10}$$

which implies $g(x) = f'(x)$, $V(x) = -f(x)f(-x)$. The most general form of the solution to (3.10) has been found by Krichever [53] in terms of the Weierstrass sigma functions which give rise to the potential (3.6). Note that (3.4) might be considered as limits of (3.5) and (3.6) as $\kappa \to 0$, and (3.6) can be regarded as double periodic form of (3.5) ((3.5) under periodic boundary conditions). The existence of M functionally independent integrals of motion in involution follows from the evident relations $d(\mathrm{tr} L^n)/dt = 0$, $1 \leq n \leq M$. The fact that all these conserved quantities are in involution also follows from the functional equation (3.10) but needs some cumbersome calculations which can be extended also to the quantum case where the time derivative should be replaced by the quantum commutator with Hamiltonian [54]. In the review paper [54], one can find a lot of interesting facts about the quantum models (3.3)–(3.6) established till 1983.

The Bethe Ansatz technique and the theory of CSM models developed independently till 1988 when Haldane [55] and Shastry [56] proposed a new spin $1/2$ model with long-range exchange resembling (1.4),

$$h(j) = J \left(\frac{\pi}{N} \right)^2 \sin^{-2} \left(\frac{\pi j}{N} \right), \tag{3.11}$$

which has very simple ground-state function of Jastrow type in the antiferromagnetic regime $J > 0$ and many degeneracies in the full spectrum. The complete integrabilty of the model and the reason of these degeneracies — the $sl(2)$ Yangian symmetry was understood later — for a comprehensive review see [57] and references therein. The Haldane–Shastry model has many nice features, including the interpretation of the excited states as ideal "spinon" gas, exact calculation of the partition function in the thermodynamic limit and the possibility of exact calculation of various correlations in the antiferromagnetic ground state [57].

The connection with the Bethe case of nearest-neighbor exchange also followed soon: in 1989, I found that the Bethe and Haldane–Shastry forms of exchange are in fact the limits of a more general model in which $h(j)$ is given by the elliptic Weierstrass function in complete analogy with (3.6),

$$h(j) = J \wp_N(j), \tag{3.12}$$

where the notation $\wp_N$ means that the real period of the Weierstrass function equals N. The absolute value of the second period π/κ is a free parameter of the model [58]. The Haldane–Shastry spin chain arises as a limit of $\kappa \to 0$. When considering the case of an infinite lattice ($N \to \infty$), one recovers the hyperbolic form of exchange (3.5) which degenerates into the nearest-neighbor exchange if $\kappa \to \infty$ under proper normalization of the coupling constant J: $J \to \kappa^{-2} \sinh^2(\kappa)J$. Hence, (3.6) might be regarded as (3.5) under periodic boundary conditions (finite lattice). Various properties of hyperbolic and elliptic spin chains form the main subject of this chapter.

The analogy between quantum spin chains and CSM models is much deeper than simple similarity of spin exchange constants and two-body CSM potentials. It concerns mainly the similarities in the form of *wave functions* of discrete and continuous cases. That is, already in [58] it has been mentioned that the solution of two-magnon problem for the exchange (3.12) and its degenerated hyperbolic form can be obtained via two-body CSM systems with potentials (3.5) and (3.6) at $l = 1$; it was also found soon to be true for three- and four-magnon wave functions for hyperbolic exchange [59]. Why does this similarity hold? Till now this is poorly understood, but it is working even for the elliptic case as it will be shown in Sections 3.3 and 3.4. Another question concerns integrability of the spin chains with hyperbolic and elliptic exchange, i.e. the existence of a family of operators commuting with the Hamiltonian. In the case of nearest-neighbor exchange, such a fam- ily can be easily found within the framework of the quantum inverse scattering method [47]. However, it is not clear up to now how this method should be used in the hyperbolic and (more general) elliptic cases. Instead, in Section 3.2 the Lax pair and empirical way of constructing conserved quantities is exposed. Section 3.5 contains various results for hyperbolic models on inhomogeneous lattices defined as equlibrium positions of the *classical* CSM hyperbolic systems in various external fields. Recent results concerning the integrability of the related Hubbard chains with variable range hopping are presented in Section 3.6. The list of still unsolved problems is given in the last Section 3.7, which contains also a short summary and discussion.

3.2. Lax Pair and Integrability

We shall consider in this section, a bit more general models with the Hamiltonian

$$\mathcal{H}_N = \frac{1}{2} \sum_{1 \leq j \neq k \leq N} h_{jk} P_{jk}, \tag{3.13}$$

where $\{P_{jk}\}$ are operators of an arbitrary representation of the permutation group S_N. The spin chains discussed above fall into this class of models, as it follows from the spin representation of the following permutation group:

$$P_{jk} = \frac{1}{2}(1 + \vec{\sigma}_j \vec{\sigma}_k).$$

The *exchange constants* h_{jk} in (3.13) are supposed to be translation invariant. The notation $\psi_{jk} = \psi(j - k)$ will be assumed for any function of the difference of numbers j and k in this section. The problem is: how to select the function h so as to get a model with integrals of motion commuting with the Hamiltonian (3.13)? The answer has been done in [58]: one can try to construct for the model the *quantum* Lax pair analogous to (3.9) and (3.10) with $N \times N$ matrices, as follows:

$$L_{jk} = (1 - \delta_{jk})f_{jk}P_{jk},$$

$$M_{jk} = (1 - \delta_{jk})g_{jk}P_{jk} - \delta_{jk}\sum_{s \neq j}^{N} h_{js}P_{js}.$$

The quantum Lax relation $[\mathcal{H}, L] = [L, M]$ is equivalent to the following functional Calogero–Moser equation for f, g, h:

$$f_{pq}g_{qr} - g_{pq}f_{qr} = f_{pr}(h_{qr} - h_{pq}), \qquad (3.14)$$

supplemented by the periodicity condition

$$h'_{pq} = h'_{p,q+N}, \qquad (3.15)$$

where $h'(x)$ is an odd function of its argument, $h'_{pq} = f_{qp}g_{pq} - f_{pq}g_{qp}$. The most general solution to (3.14) has been given in [53] as the combination of the Weierstrass sigma functions. There is the normalization of f and h which allows one to write the relations

$$h(x) = f(x)f(-x), \quad g(x) = \frac{df(x)}{dx}, \quad h'(x) = \frac{dh(x)}{dx}.$$

The solution given in [53] looks as

$$f(x) = \frac{\sigma(x + \alpha)}{\sigma(x)\sigma(\alpha)}\exp(-x\zeta(\alpha)), \quad h(x) = \wp(\alpha) - \wp(x). \qquad (3.16)$$

where

$$\zeta'(x) = -\wp(x), \quad \frac{d(\log\sigma(x))}{dx} = \zeta(x).$$

and α is the spectral parameter which does not introduce any new in exchange dynamics. This form of the solution can be easily verified with the use of the identity

$$\zeta(x+y+z) - \zeta(x) - \zeta(y) - \zeta(z) = \frac{\sigma(x+y)\sigma(x+z)\sigma(y+z)}{\sigma(x+y+z)\sigma(x)\sigma(y)\sigma(z)}.$$

The periodicity condition (3.15) means that all Weierstrass functions in (3.16) are defined on the torus $\mathbf{T}_N = \mathbf{C}/N\mathbf{Z} + i\frac{\pi}{\kappa}\mathbf{Z}$, $\kappa \in \mathbf{R}_+$ is the free parameter of the model. It is easy to see that the exchange (3.16) reduces in the limit $\kappa \to 0$ to the Haldane–Shastry model and the limit of infinite lattice size ($N \to \infty$) corresponds to the hyperbolic variable range form of exchange. And finally, in the limit $\kappa \to \infty$, just nearest-neighbor exchange (3.2) is reproduced as it was already mentioned in the preceding section.

However, the problem is not a classical one and the existence of the Lax representation does not guarantee the existence of the integrals of motion as invariants of the L matrix. In fact, just for the problem under consideration the operators trL^n do *not* commute with the Hamiltonian. Nevertheless, already in [58] the way of constructing integrals of motion on the base of f function of Lax pair was proposed. That is, it was found that the operator

$$J(\alpha) = \sum_{j \neq k \neq l} f_{jk}f_{kl}f_{lj}P_{jk}P_{kl} \tag{3.17}$$

commutes with $\mathcal{H}$! Moreover, the dependence of the right-hand side on the spectral parameter implies that there are two functionally independent operators bilinear in $\{P\}$ and commuting with $\mathcal{H}$,

$$J_1 = \sum_{j \neq k \neq l} (\zeta(j-k) + \zeta(k-l) + \zeta(l-j))P_{jk}P_{kl},$$

$$J_2 = \sum_{j \neq k \neq l} [2(\zeta(j-k) + \zeta(k-l) + \zeta(l-j))^3$$
$$+ \wp'(j-k) + \wp'(k-l) + \wp'(l-j)]P_{jk}P_{kl}.$$

Very long but straightforward calculations show that $J_{1,2}$ mutually commute.

It turns out [73] that the construction (3.17) can be generalized for more complicated operators with higher degrees of $\{P\}$. The basic idea is to use the operators of cyclic permutations $P_{s_1,\ldots,s_l} \equiv P_{s_1 s_2} P_{s_2 s_3,\ldots,} P_{s_{l-1} s_l}$ and functions $F_{s_1,\ldots,s_l} = f_{s_1 s_2} f_{s_2 s_3,\ldots,} f_{s_{l-1} s_l} f_{s_l s_1}$ which are invariant under the action of elements of a group of cyclic permutations of subindices $(1,\ldots,l)$. If one denotes as $\Phi(s_1,\ldots,s_l)$ the functions which are completely symmetric in their arguments, and $\sum_{C \in C_l} B_{s_1,\ldots,s_l}$ as the sum over all cyclic permutations of the subindices of $B_{s_1,\ldots,sl}$, the following properties of the above objects are useful:

(A) *The functions $h(x)$ and $h'(x)$ obey the relation*

$$\sum_{C \in C_3} h'_{s_1 s_2}(h_{s_1 s_3} - h_{s_2 s_3}) = 0.$$

(B) *The sum $F^{(C)}_{s_1 \ldots s_{l+1}} = \sum_{C \in C_l} F_{s_1,\ldots,s_l s_{l+1}}$ does not depend on s_{l+1}.*

(C) *The sum $\sum_{s_1 \neq \ldots s_{l+1}} \Phi(s_1,\ldots,s_{l+1}) F_{s_1,\ldots,s_l}(h_{s_{l+1} sl} - h_{s_{l+1} s_1}) P_{s_1 \ldots s_{l+1}}$ vanishes for any symmetric function Φ.*

(D) *The sum*

$$S_l(\Phi) = \sum_{s_1 \neq \ldots s_l} F_{s_1,\ldots,s_l} \Phi(s_1,\ldots,s_l)(h_{s_l s_{l-1}} - h_{s_1 s_l}) P_{s_1,\ldots,s_{l-1}}$$

has a representation in the form $S^{(1)}_{l-1}(\Phi) + S^{(2)}_{l-1}(\Phi)$, where

$$S^{(1)}_{l-1}(\Phi) = \sum_{s_1 \neq \ldots s_{l-1}} F_{s_1 \ldots s_{l-1}}$$

$$\times \left(\frac{1}{l-1} \sum_{p \neq s_1,\ldots,s_{l-1}}^{N} \Phi(s_1,\ldots,s_{l-1},p) \sum_{v=1}^{l-1} h'_{ps_v} \right) P_{s_1 \ldots s_{l-1}},$$

and

$$S^{(2)}_{l-1}(\Phi) = \sum_{s_1 \neq \ldots s_{l-1}} F_{s_1 \ldots s_{l-2}}(h_{s_{l-1} s_{l-2}} - h_{s_{l-1} s_l})$$

$$\times \left(\sum_{p \neq s_1 \ldots, s_{l-1}} h_{s_{l-1} p} \Phi(s_1,\ldots,s_{l-1},p) \right) P_{s_1 \ldots s_{l-1}}$$

if l > 3

$$S_2^{(2)}(\Phi) = -\frac{1}{2} \sum_{s_1 \neq s_2} h'_{s_1 s_2} \sum_{p \neq s_1, s_2}^{N} (h_{s_1 p} - h_{s_2 p}) \Phi(s_1, s_2, p) P_{s_1 s_2}.$$

The main statement concerning the integrals of motion for the Hamiltonian (3.13) can be proved without the use of the specific form (3.16) of the solution to the Calogero–Moser equation. It can be formulated as follows: Let $I_m (3 \leq m \leq N)$ be the linear combinations of the operators of cyclic permutations in ordered sequences of N symbols,

$$I_m = \sum_{l=0}^{[\frac{m}{2}]-1} \frac{(-1)^l}{m - 2l} \sum_{s_1 \neq \ldots s_{m-2l}} \Phi^{(l)}(s_1, \ldots, s_{m-2l}) F_{s_1 \ldots s_{m-2l}} P_{s_1 \ldots s_{m-2l}}.$$

(3.18)

Then they will give the integrals of motion as it follows from

Proposition 2.1. *The operators I_m commute with $\mathcal{H}_N$ given by (3.1) if the functions $\Phi^{(l)}$ are determined by the recurrence relation*

$$\Phi^{(0)} = 1, \quad \Phi^{(l)}(s_1, \ldots, s_{m-2l}) = l^{-1}$$

$$\times \sum_{l \leq j < k \leq N; j, k \neq s_1, \ldots, s_m - 2l} h_{jk} \Phi^{(l-1)}(s_1, \ldots, s_{m-2l}, j, k) \quad (3.19)$$

or, equivalently, are given by sums over 2l indices

$$\Phi^{(l)}(s_1, \ldots, s_{m-2l}) = (l!)^{-1}$$

$$\times \sum_{1 \leq j_\alpha < k_\alpha \leq N; \{j, k\} \neq s_1, \ldots, s_m - 2l} \lambda_{\{jk\}} \prod_{\alpha=1}^{l} h_{j_\alpha k_\alpha}, \quad (3.20)$$

where $\lambda_{\{jk\}}$ equals 1 if the product $\prod_{\alpha \neq \beta}^{l} (j_\alpha - j_\beta)(k_\alpha - k_\beta)(j_\alpha - k_\beta)$ differs from zero and vanishes otherwise.

The rigorous proof of the statements A–D can be found in [73]. Here, we give only a sketch of the proof of Proposition 2.1. It is based on the calculation of the commutator

$$J_n = \sum_{s_1 \neq \ldots, s_n} [\Phi(s_1, \ldots, s_n) F_{s_1, \ldots, s_n} P_{s_1, \ldots, s_n}, \mathcal{H}_N]. \quad (3.21)$$

where Φ is symmetric in its variables. With the use of invariance of $F_{s_1,\ldots,s_n}$ and $P_{s_1,\ldots,s_n}$ under cyclic changes of summation variables, it is easy to show that this commutator can be written as

$$J_n = n \left[J_n^{(1)} + J_n^{(2)} + \sum_{\nu=2}^{[n/2]} \left(1 - \left(\frac{n-1}{2} - \left[\frac{n}{m} \right] \right) \delta_{\nu,[n/2]} \right) J_{n,\nu}^{(3)} \right],$$

where

$$J_n^{(1)} = \sum_{s_1 \neq,\ldots,s_{n+1}} \Phi(s_1,\ldots,s_n) F_{s_1,\ldots,s_n} (h_{s_n s_{n+1}} - h_{s_1 s_{n+1}}) P_{s_1,\ldots,s_{n+1}},$$

$$J_n^{(2)} = \sum_{s_1 \neq,\ldots,s_n} \Phi(s_1,\ldots,s_n) F_{s_1,\ldots,s_n} (h_{s_{n-1} s_n} - h_{s_1 s_n}) P_{s_1,\ldots,s_{n-1}},$$

$$J_{n,\nu}^{(3)} = \sum_{s_1 \neq,\ldots,s_n} \Phi(s_1,\ldots,s_n) F_{s_1,\ldots,s_n} (h_{s_\nu s_n} - h_{s_1 s_{\nu+1}})$$
$$\times P_{s_1,\ldots,s_\nu} P_{s_{\nu+1},\ldots,s_n}.$$

The third term can be transformed with the use of the functional equation (3.14) and cyclic symmetry of $P_{s_1,\ldots,s_\nu}$ and $P_{s_{\nu+1},\ldots,s_n}$ to the form

$$J_{n,\nu}^{(3)} = \sum_{s_1 \neq \ldots s_n} \Phi(s_1,\ldots,s_n)[\nu^{-1} \varphi_{s_{\nu+1},\ldots,s_{n-1} s_n}$$
$$\times (F_{s_1,\ldots,s_\nu s_{\nu+1}}^{(C)} - F_{s_1,\ldots,s_\nu s_n}^{(c)}) + F_{s_{\nu+1},\ldots,s_n}$$
$$\times (\varphi_{s_{\nu+1} s_1,\ldots,s_\nu} - \varphi_{s_1,\ldots,s_{\nu+1}})]P_{s_1,\ldots,s_\nu} P_{s_{\nu+1},\ldots,s_n}, \quad (3.22)$$

where $\varphi_{s_1 s_2,\ldots,s_{l+1}} = f_{s_1 s_2} f_{s_2 s_3},\ldots,f_{s_l s_{l+1}} g_{s_{l+1} s_1}$.

Now it is easy to see that the term in the first brackets in (3.22) disappears due to statement (B) and the term in the second brackets vanishes due to the relation

$$F_{s_1,\ldots,s_l}(h_{s_1 s_{l+1}} - h_{s_l s_{l+1}}) = \varphi_{s_1 s_2,\ldots,s_{l+1}} - \varphi_{s_{l+1} s_1,\ldots,s_l},$$

which allows to transform this term to the expression which vanishes upon symmetrization in all cyclic changes of $s_1,\ldots,s_\nu$. Hence, the operator (3.22) contains only cyclic permutations of rank $(n + 1)$ and $(n - 1)$. This fact leads to the idea of recurrence construction of the operators (3.18) which would commute with the Hamiltonian. It happens if the functions $\{\Phi\}$ obey the recurrence relation

$$\sum_{p \neq s_1, \ldots, s_{m-2l-1}} (h_{p s_{m-2l-1}} - h_{p s_1}) \Phi^{(l)}(s_1, \ldots, s_{m-2l-1}, p)$$

$$= \Phi^{(l+1)}(s_1, s_2, \ldots, s_{m-2l-2}) - \Phi^{(l+1)}$$

$$\times (s_{m-2l-1}, s_2, \ldots, s_{m-2l-2}), \Phi^{(0)} = 1,$$

which can be solved in the form (3.19) or (3.20).

The dependence of (3.18)–(3.20) on the spectral parameter α via the relations (3.16) allows one to conclude that there are several integrals of motion at each m. That is, the analysis of this dependence shows that the operators (3.18) can be written in the form

$$I_m = w_m(\alpha) P_m + \sum_{\mu=1}^{m-2} w_{m-\mu}(\alpha) I_{m,\mu} + I_{m,m},$$

where P_m commutes with all operators of elementary transpositions, $I_{m,\mu}$ are linear combinations of $\{P_{s_1,\ldots,s_{m-2l}}\}$ which do not depend on α, and $w_{m-\mu}(\alpha)$ are linearly independent elliptic functions of the spectral parameter. Nothing is known for the mutual commutativity of these operators except the explicit result for $m = 3$. In the last case, the proof of the commutativity is as follows.

Due to the arbitrariness of the spectral parameter α, (3.17) in fact contains only three independent integrals of motion,

$$J = -\frac{1}{2} \wp'(\alpha) J_0 + \wp(\alpha) J_1 - \frac{1}{2} J_2,$$

$$J_0 = \sum_{j \neq k \neq l}^{N} P_{jkl},$$

$$J_1 = \sum_{j \neq k \neq l}^{N} \varphi_{jkl} P_{jkl}, \quad \varphi_{jkl} = \zeta(j - k) + \zeta(k - l) + \zeta(l - j), \quad (3.23)$$

$$J_2 = \sum_{j \neq k \neq l}^{N} F_{jkl} P_{jkl}, \qquad (3.24)$$

$$F_{jkl} = \frac{1}{3}\{2[\zeta(j-k) + \zeta(k-l) + \zeta(l-j)][\wp(j-k) + \wp(k-l)$$
$$+ \wp(l-j)] + \wp'(j-k) + \wp'(k-l) + \wp'(l-j)\}. \qquad (3.25)$$

The integral of motion J_0, related to the total spin for the case of spin chains, trivially commutes with any transposition P_{jk} and therefore also with H, J_1 and J_2. It is easy to prove the identities

$$2[\zeta(j-k) + \zeta(k-l) + \zeta(l-j)]\wp'(j-k) + \wp'(j-k)$$
$$= 2[\zeta(j-k) + \zeta(k-l) + \zeta(l-j)]$$
$$\times \wp(k-l) + \wp'(k-l) = 2[\zeta(j-k) + \zeta(k-l)$$
$$+ \zeta(l-j)]\wp(l-j) + \wp'(l-j),$$

which allow one to rewrite F_{jkl} in one of the following forms:

$$F_{jkl} = 2[\zeta(j-k) + \zeta(k-l) + \zeta(l-j)]\wp(j-k) + \wp'(j-k)$$
$$= 2[\zeta(j-k) + \zeta(k-l) + \zeta(l-j)]\wp(l-j) + \wp'(l-j). \qquad (3.26)$$

Note that both φ_{jkl} and F_{jkl} are antisymmetric with respect to permutations of their indices.

The commutator of the operators (3.23), (3.24) can be written down in the form

$$[J_1, J_2] = \sum_{\substack{j,k,l=1 \\ j \neq k \neq l \neq j}}^{N} \sum_{\substack{m,n,p=1 \\ m \neq n \neq p \neq m}}^{N} \varphi_{jkl} F_{mnp} [P_{jkl}, P_{mnp}], \qquad (3.27)$$

The commutator at the right-hand side of (3.27) might be non-zero if and only if one or two indices (mnp) coincide with (jkl). Consider first the coincidence of one index (say, m) with one of (jkl). The direct calculation of this contribution to the commutator can be written as

$$J_3 = 9 \sum_{\substack{j,k,l,n,p=1 \\ \text{all different}}}^{N} (\varphi_{jnp} F_{jkl} - \varphi_{jkl} F_{jnp}) P_{jklnp},$$

where $P_{jklnp} = P_{jk} P_{kl} P_{ln} P_{np}$ is symmetric with respect to all cyclic permutations of its indices. Hence, the coefficient in front of it can be rewritten due to this symmetry, and one finds

$$J_3 = \frac{9}{5} \sum_{\substack{j,k,l,n,p=1 \\ \text{all different}}}^{N} \Omega_{jklnp} P_{jklnp},$$

where

$$\Omega_{jklnp} = F_{jkl}(\varphi_{jnp} - \varphi_{lnp}) + F_{jnp}(\varphi_{nkl} - \varphi_{jkl})$$
$$+ F_{kln}(\varphi_{jkp} - \varphi_{jnp}) + F_{jkp}(\varphi_{pln} - \varphi_{kln})$$
$$+ F_{lnp}(\varphi_{jkl} - \varphi_{jkp}). \tag{3.28}$$

The function Ω_{jklnp} in fact depends on four arguments due to the fact that φ and F depend only on differences of their indices. Let us introduce the notation

$$p - j = x, p - k = y, p - l = z, n - p = v.$$

Then other differences can be written as

$$n - j = v + x, n - k = v + y, n - l = v + z,$$
$$j - k = y - x, j - l = z - x, k - l = z - y.$$

With the use of the above formulas, one can rewrite Ω (3.28) as

$$\Omega_{jklnp} = R(x, y, z, v),$$

where

$$R(x, y, z, v)$$
$$= [2(\zeta(y - x) + \zeta(z - y) + \zeta(x - z))\wp(x - z) + \wp'(x - z)]$$
$$\times [-\zeta(v + x) + \zeta(x) + \zeta(v + z) - \zeta(z)]$$
$$+ [2(\zeta(x) + \zeta(y - x) - \zeta(y))\wp(y) - \wp'(y)][\zeta(z) + \zeta(v)$$
$$- \zeta(z - y) - \zeta(v + y)]$$

$$+ [2(\zeta(v) + \zeta(x) - \zeta(v + x))\wp(v + x) - \wp'(v + x)]$$
$$\times [\zeta(v + y) - \zeta(v + z) - \zeta(y - x) + \zeta(z - x)]$$
$$+ [2(-\zeta(v + z) + \zeta(v) + \zeta(z))\wp(z) + \wp'(z)]$$
$$\times [\zeta(x - z) + \zeta(z - y) - \zeta(x) + \zeta(y)]$$
$$+ [2(\zeta(z - y) - \zeta(v + z) + \zeta(v + y))\wp(v + y)$$
$$+ \wp'(v + y)][\zeta(y - x) - \zeta(y) - \zeta(v) + \zeta(v + x)]. \qquad (3.29)$$

The goal is now to simplify this very cumbersome formula. First, let us note that $R(x, y, z, v)$ is elliptic, i.e. double periodic function of all its arguments. And second, I shall use the following Laurent decomposition of $\zeta(x)$ and $\wp(x)$ near $x = 0$, their only singularity point,

$$\wp(x) \sim x^{-2} + ax^2 + O(x^4), \quad \zeta(x) \sim x^{-1} - \frac{a}{3} + O(x^5),$$

and the differential equations for the Weierstrass $\wp$ function,

$$\wp'(x)^2 = 4\wp(x)^3 - g_2\wp(x) - g^3, \quad \wp''(x) = 6\wp(x)^2 - \frac{g_2}{2},$$

where $a = \frac{g_2}{20}, g_2, g_3$ are some constants.

Consider now $R(x, y, z, v)$ as the elliptic function of v. It can have simple poles at four points: $v = 0$, $v = -x$, $v = -y$, $v = -z$ and no other singularities on the torus $\mathbf{T} = \mathbf{C}/(\mathbf{Z}N + \mathbf{Z}\omega)$. It might be equal to zero if we would prove that all these poles are in fact absent (in this case R does not depend on v), and that $R(x, y, z, 0) = 0$.

Calculating the Laurent decomposition of R near the point $v = 0$ gives

$$R(x, y, z, v) \sim v^{-1}A(x, y, z) + B(x, y, z) + \cdots,$$

where

$$A(x, y, z) = 2[\wp(x)(\zeta(y) - \zeta(z) - \zeta(y - x) + \zeta(z - x))$$
$$+ \wp(z)(\zeta(x - z) + \zeta(z - y) - \zeta(x) + \zeta(y)) + \wp(y)(\zeta(x)$$
$$+ \zeta(z) - 2\zeta(y) + \zeta(y - x) + \zeta(y - z)) - \wp'(y)],$$

$$B(x, y, z) = -\wp'(x)(\zeta(z) - \zeta(y) + \zeta(y - x) - \zeta(z - x))$$
$$+ \wp'(z)(\zeta(y) - \zeta(x) + \zeta(x - z) + \zeta(z - y))$$
$$+ \wp'(y)(\zeta(x) + \zeta(z) - 2\zeta(y) - \zeta(z - y) + \zeta(y - x))$$
$$+ 2(\wp(x) - \wp(y))(\wp(z) - \wp(y)) - \wp''(y).$$

Consider $A(x, y, z)$ as the elliptic function of the argument x. It might have poles at $x = 0$, $x = y$, $x = z$. It is easy to calculate the first two terms of its Laurent expansion near $x = 0$:

$$A(x, y, z) \sim 2\Big\{ x^{-2}[x(\wp(z) - \wp(y)) + \frac{x^2}{2}(\wp'(y) - \wp'(z))]$$
$$+ \wp(z)(-x^{-1} + \zeta(y) + \zeta(z - y) - \zeta(z)) + \wp(y)(x^{-1} + \zeta(z) - \zeta(y)$$
$$+ \zeta(y - z)) - \wp'(y) \Big\} = -(\wp'(y) + \wp'(z)) + 2(\wp(y)$$
$$- \wp(z))(\zeta(z) - \zeta(y) + \zeta(y - z)).$$

Now one sees that $A(x, y, z)$ has no pole at $x = 0$ and $A(0, y, z) = 0$ due to the known identity (change z to $-z$ in the composition formula for ζ)

$$\frac{1}{2}(\wp'(y) + \wp'(z)) = (\wp(y) - \wp(z))(\zeta(z) - \zeta(y) + \zeta(y - z)).$$
$$(3.30)$$

An easy calculation shows that there are no poles of A at $x = y$ and $x = z$. One concludes that

$$A(x, y, z) \equiv 0.$$

Let us now simplify the expression for $B(x, y, z)$. It is easy to see that at $x = y$ and $x = z$ there are no poles of this function. The calculation shows that there is no pole at $x = 0$ too and gives

$$B(0, y, z) = \frac{1}{3}(\wp''(y) - \wp''(z)) + (\wp'(z) - \wp'(y))(\zeta(y) - \zeta(z)$$
$$+ \zeta(z - y)) - 2\wp(y)(\wp(z) - \wp(y)) - \wp''(y).$$

By using the identity (3.30) and differential equations for the Weierstrass function, one can write $B(0, y, z)$ in the form

$$B(0, y, z) = 2(\wp(y)^2 - \wp(z)^2) - 2\wp(y)(\wp(z) - \wp(y))$$
$$+ \frac{1}{2} \frac{\wp'(z)^2 - \wp'(y)^2}{\wp(z) - \wp(y)} - 6\wp(y)^2 + \frac{g_2}{2} = 0.$$

Hence, the elliptic function $B(x, y, z)$ has no poles and $B(0, y, z) = 0$. It results in the identity

$$B(x, y, z) \equiv 0.$$

Let us summarize these steps of calculations. It is proved that $R(x, y, z, v)$ has no pole at $v = 0$ and $R(x, y, z, 0) = 0$. But it might have poles at $v = -x, -y, -z$. Calculation of the asymptotics at $v \to -x$ gives

$$R(x, y, z, v) \sim \frac{1}{v + x}[2(\zeta(z - y) + \zeta(x - z) + \zeta(y - x))$$
$$\times (\wp(y - x) - \wp(x - z)) + \wp'(y - x) + \wp'(z - x)].$$

But the identity (3.30) shows that the right-hand side of the above formula is just zero. Similar calculations result in the absence of poles of $R(x, y, z, v)$ at $v = -y$ and $v = -z$. Hence, this function has no poles in v at all and $R(x, y, z, 0) = 0$. One comes up to the identity

$$R(x, y, z, v) \equiv 0. \tag{3.31}$$

It means that all contributions to the commutator $[J_1, J_2]$ quartic in permutation operators disappear. Let us consider now the case of coinciding two pairs of indices in the sets (jkl), (mnp) in (3.27). The corresponding contribution to the commutator consists of two parts,

$$J_4 = 9 \sum_{\substack{j,k,l,n=1 \\ \text{all different}}}^{N} (\varphi_{jkl} F_{jkn} - \varphi_{jkn} F_{jkl}) P_{jl} P_{kn}, \tag{3.32}$$

$$J_5 = 9 \sum_{\substack{j,k,l,p=1 \\ \text{all different}}}^{N} F_{jkp}(\varphi_{ljp} - \varphi_{klp}) P_{jkl}. \tag{3.33}$$

The operator in (3.32) is invariant under changing indices $(j \leftrightarrow l)$; $(k \leftrightarrow n)$; $(j \leftrightarrow k, l \leftrightarrow n)$. Symmetrization of the coefficient in front of it gives, after an easy calculation taking into account the antisymmetry of φ_{jkl} and F_{jkl} under the transposition of two indices, that $J_4 \equiv 0$ for otherwise arbitrary φ_{jkl}, F_{jkl}.

It remains to calculate J_5. Since P_{jkl} is symmetric with respect to the cyclic permutations of (jkl), (3.32) can be written in the form

$$J_5 = 3 \sum_{\substack{j,k,l,p=1 \\ \text{all different}}} T_{jklp} P_{jkl},$$

$$T_{jklp} = F_{jkp}(\varphi_{ljp} - \varphi_{klp}) + F_{ljp}(\varphi_{klp} - \varphi_{jkp}) + F_{klp}(\varphi_{jkp} - \varphi_{ljp}).$$
$$(3.34)$$

It is of use to introduce the notation

$$j - p = x, k - p = y, l - p = z.$$

Then

$$j - k = x - y, k - l = y - z, j - l = x - z,$$

and one can rewrite (3.34) with the use of (3.23) and (3.25) as

$$
\begin{aligned}
T_{jklp} &= \Phi(x, y, z) \\
&= [2(\zeta(x-y) + \zeta(y) - \zeta(x))\wp(x-y) + \wp'(x-y)] \\
&\quad \times [\zeta(z-x) + \zeta(z-y) + \zeta(x) + \zeta(y) - 2\zeta(z)] \\
&\quad + [2(\zeta(z-x) + \zeta(x) - \zeta(z))\wp(z-x) + \wp'(z-x)] \\
&\quad \times [\zeta(y-z) + \zeta(y-x) + \zeta(x) + \zeta(z) - 2\zeta(y)] \\
&\quad + [2(\zeta(y-z) + \zeta(z) - \zeta(y))\wp(y-z) + \wp'(y-z)] \\
&\quad \times [\zeta(x-y) + \zeta(x-z) + \zeta(y) + \zeta(z) - 2\zeta(x)].
\end{aligned}
$$

It is easy to see that $\Phi(x, y, z)$ is antisymmetric with respect to permutations of its arguments,

$$\Phi(x, y, z) = -\Phi(y, x, z) = -\Phi(x, z, y).$$

The problem consists now in simplifying $\Phi(x, y, z)$ which is an elliptic function of all its arguments. As a function of x, it has poles at $x = 0$,

$x = y$, $x = z$. The calculation of the first two terms of its Laurent expansion near $x = 0$ gives

$$\begin{aligned}
\Phi(x, y, z) \sim{}& 2x^{-2}[\wp(z) - \wp(y)] + x^{-1} \\
&\times \{2[\zeta(z - y) - \zeta(z) + \zeta(y)] \\
&\times [2\wp(y - z) - \wp(y) - \wp(z)] + \wp'(y) \\
&- \wp'(z) - 2\wp'(y - z)\}.
\end{aligned}$$

The coefficient at x^{-1} can be drastically simplified by using the identity (3.30). Implying it two times results in

$$\Phi(x, y, z) \sim 2x^{-2}[\wp(z) - \wp(y)] + 2x^{-1}[\wp'(y) - \wp'(z)]. \quad (3.35)$$

The first coefficients in the Laurent expansions near the points $x = y$ and $x = z$ are

$$\Phi(x, y, z) \sim -2(x - y)^{-1}\wp'(y), \quad \Phi(x, y, z) \sim 2(x - z)^{-1}\wp'(z). \quad (3.36)$$

Consider now the trial function

$$\Psi(x, y, z) = 2\{\wp(x - y)[\wp(y) - \wp(x)] + \wp(x - z)[\wp(x) - \wp(z)]\}.$$

It is easy to see that it has poles at $x = 0$, $x = y$, $x = z$ with the same residues (3.35) and (3.36) as $\Phi(x, y, z)$. Hence,

$$\Phi(x, y, z) = \Psi(x, y, z) + \psi(y, z),$$

where $\psi(y, z)$ does not depend on x. Now, with the use of antisymmetry of Φ, one finds that the only choice for ψ is

$$\psi(y, z) = 2\wp(y - z)[\wp(z) - \wp(y)],$$

and finally one can write the remarkable identity

$$\begin{aligned}
\Phi(x, y, z) = {}& 2\{\wp(x - y)[\wp(y) - \wp(x)] + \wp(x - z) \\
&\times [\wp(x) - \wp(z)] + \wp(y - z)[\wp(z) - \wp(y)]\}. \quad (3.37)
\end{aligned}$$

Now it is possible to prove the relation

$$\sum_{\substack{p=1 \\ p \neq j,k,l}}^{N} \Phi(x, y, z) = 0, \tag{3.38}$$

for any fixed $j \neq k \neq l \neq j$. Indeed, coming back to the notation of the arguments of Φ and using (3.37), one finds

$$\sum_{\substack{p=1 \\ p \neq j,k,l}}^{N} \Phi(x, y, z) = 2\{\wp(j - k)[q(k, l, j) - q(j, k, l)]$$

$$+ \wp(j - l)[q(j, k, l) - q(l, j, k)]$$

$$+ \wp(l - k)[q(l, j, k) - q(k, l, j)]\},$$

where

$$q(k, l, j) = \sum_{\substack{p=1 \\ p \neq j,k,l}}^{N} \wp(k - p) = S(k) - \wp(k - j) - \wp(k - l),$$

$$S(k) = \sum_{\substack{p=1 \\ p \neq k}}^{N} \wp(k - p).$$

But $S(k)$ does not depend on k since $\wp(k - p)$ is periodic with the period N. Now it is easy to see that (3.38) holds for all j, k, l and the commutator $[J_1, J_2]$ vanishes.

One can prove now that the integrals of motion J_0, J_1 and J_2 are linearly independent for $N > 4$. More specifically, one can prove that the operator J_0 is linearly independent of J_1 and J_2 for $N \geq 3$, operators J_1 and J_2 are linearly dependent for $N = 3, 4$, and operators J_1, J_2 are linearly independent for $N > 4$.

To study the linear independence, one looks for the complex numbers λ, μ, ρ such that

$$\lambda J_0 + \mu J_1 + \rho J_2 = 0.$$

As the coefficients in Equations (3.23) and (3.25) are symmetrized with respect to the cyclic permutations of indices, the last relation is equivalent to

$$\lambda + \mu\varphi_{jkl} + \rho F_{jkl} = 0$$

for any mutually different $j, k, l = 1, \ldots, N$. As φ_{jkl} and F_{jkl} are antisymmetric under the exchange of two indices, this is further equivalent to

$$\lambda = 0, \quad \mu\varphi_{jkl} + \rho F_{jkl} = 0.$$

In particular, J_0 is linearly independent of J_1 and J_2.

Consider now the case of $N = 3$. Here,

$$J_1 = 3\varphi_{123}(J_{123} - J_{213}), \quad J_2 = 3F_{123}(J_{123} - J_{213})$$

so J_1 and J_2 are linearly dependent for $N = 3$. In the case of $N = 4$, one obtains remembering that N is the period of Weierstrass functions in these considerations and their other properties

$$J_1 = 3\varphi_{123}(P_{123} - P_{213} + P_{124} - P_{214} + P_134 - P_{314} + P_{234} - P_{324}),$$

$$J_2 = 3F_{123}(P_{123} - P_{213} + P_{124} - P_{214} + P_{134} - P_{314} + P_{234} - P_{324}),$$

with

$$\varphi_{123} = \zeta(2) - 2\zeta(1), \quad F_{123} = \frac{2}{3}[2\varphi_{123}^3 - \wp'(1)],$$

and the linear dependence of J_1 and J_2 is seen for $N = 4$.

Let us further on assume $N > 4$ and assume that there exists μ and ρ satisfying equations $\mu\varphi_{jkl} + \rho F_{jkl} = 0$ for every possible j, k, l. Let us fix k and l and define function

$$\psi(z) = \mu a(z) + \rho b(z),$$

$$a(z) = \zeta(z - k) + \zeta(k - l) + \zeta(l - z),$$

$$c(z) = \wp'(z - k) + \wp'(k - l) + \wp'(l - z),$$

$$b(z) = \frac{1}{3}[c(z) + 2a(z)^3],$$

such that our equations read $\psi(j) = 0$ for $j \in \{1, \ldots, N\}\backslash\{k, l\}$. ψ is an elliptic function with periods N and ω. The only possible poles are at the points $z = k$ and $z = l$. They are simple poles of a, let us calculate the behavior of $b(z)$ for $z = k + x$, $x \to 0$:

$$a(z) = \frac{1}{x} - \zeta'(l - k)x + O(x^2) = \frac{1}{x} + \wp(l - k)x + O(x^2),$$

$$c(z) = -\frac{2}{x^3} + O(x),$$

$$b(z) = \frac{2\wp(l - k)}{x} + O(1).$$

Similar formulas hold for $z \to l$ due to the antisymmetry of a, c, b with respect to the interchange of k and l. Therefore, ψ has at most simple poles at k and l. By Liouville theorem, ψ can have at most two zeroes (modulo periods) if it is not constant. However, there are at least $N - 2 > 2$ zeros at the points $z = j$. So $\psi(z) \equiv 0$. Looking for the behavior at $z \to k$, one sees that equation

$$\mu + 2\wp(l - k)\rho = 0$$

must be valid. The function $\wp$ can take the same value at most twice (modulo periods) due to the Liouville theorem. As $k \neq l$ can be chosen arbitrarily, one finds two different values among $N - 1$ numbers $\wp(l - k)$ for $l - k = 1, \ldots, N - 1$. So necessarily $\mu = \rho = 0$ and the linear independence of J_0, J_1, J_2 is proved for $N > 4$. Their linear independence of H is trivial as different permutations enter the definition of H.

To summarize, it was proved that the Hamiltonian (3.13) and operators J_0, J_1 (3.23) and J_2 (3.25) are linearly independent and generate the commutative ring. As a byproduct, one obtains the remarkable identities between elliptic functions (3.29), (3.31), and (3.37). The proof was based on direct evaluation of $[J_1, J_2]$ due to the lack of any other methods. The model (3.1) with elliptic form of $h(j - k)$ is still not immersed in the scheme of the quantum inverse scattering method.

Still there is no connection with Yang–Baxter theory, i.e. the corresponding R-matrix and L-operators are unknown. There is the excellent paper by Hasegawa [74] which states that R-matrix for spinless elliptic quantum Calogero–Moser systems is Belavin's, and also of elliptic type. However, it is not clear how to extend Hasegawa's method to the spin case so as to reproduce the rich variety of the operators (3.18).

3.3. The Infinite Chain

On the infinite line, the model is defined by the Hamiltonian

$$H = -\frac{J}{2} \sum_{j \neq k} \frac{\kappa^2}{\sinh^2 \kappa(j-k)} (\vec{\sigma}_j \vec{\sigma}_k - 1)/2, \qquad (3.39)$$

where $j, k \in \mathbf{Z}$. At these conditions, only the ferromagnetic case $J > 0$ is well defined. The spectrum to be found consists of excitations over ferromagnetic ground state $|0 >$ with all spins up, which has zero energy. The energy of one spin wave is just given by Fourier transform of the exchange in (3.39),

$$\varepsilon(p) = J \left\{ -\frac{1}{2} \wp_1 \left(\frac{ip}{2\kappa} \right) + \frac{1}{2} \left[\frac{p}{\pi} \zeta_1 \left(\frac{i\pi}{2\kappa} \right) \right.\right.$$

$$\left.\left. - \zeta_1 \left(\frac{ip}{2\kappa} \right) \right]^2 - \frac{2i\kappa}{\pi} \zeta_1 \left(\frac{i\pi}{2\kappa} \right) \right\}, \qquad (3.40)$$

where Weierstrass functions $\wp_1$, ζ_1 are defined on the torus $\mathbf{T}_1 = \mathbf{C}/(\mathbf{Z} + \frac{i\pi}{\kappa}\mathbf{Z})$, i.e. $\wp_1$ has the periods $(1, \omega = i\pi/\kappa)$.

3.3.1. *Two-magnon scattering*

The two-magnon problem for the model (3.39) is already non-trivial. One has to solve the difference equation for two-magnon wave function $\psi(n_1, n_2)$, which is defined by the relation

$$|\psi\rangle = \sum_{n_1 \neq n_2} \psi(n_1, n_2) s_{n_1}^- s_{n_2}^- |0\rangle, \qquad (3.40)$$

where the operator $\{s_{n_\alpha}^-\}$ reverses spin at the site n_α and $|\psi\rangle$ is an eigenvector of the Hamiltonian (3.39). The solution is based on the formula [58]

$$\sum_{k=-\infty}^{\infty} \frac{\kappa^2 \exp(ikp)}{\sinh^2[\kappa(k+z)]} \coth\kappa(k+l+z)$$

$$= -\frac{\sigma_1(z+r_p)}{\sigma_1(z-r_p)} \coth(\kappa l) \exp\left[\frac{pz}{\pi}\zeta_1(\omega/2)\right]$$

$$\times \left\{ \wp_1(z) - \wp_1(r_p) + 2\left[\zeta_1(r_p) - \frac{2r_p}{\omega}\zeta\left(\frac{\omega}{2}\right)\right.\right.$$

$$\left. + \frac{\kappa}{\sinh(2\kappa l)}(1-\exp(-ipl))\right]$$

$$\times (\zeta_1(z+r_p) - \zeta_1(z) + 2\zeta_1(r_p) - \zeta_1(2r_p)) \Bigg\} \qquad (3.41)$$

where $r_p = -\omega p/4\pi$ and $l \in \mathbf{Z}$.

The proof of (3.41) is based on the quasiperiodicity of the sum on its left-hand side and the structure of its only singularity at the point $z = 0$ on a torus $\mathbf{T}_1$ obtained by factorization of a complex plane on the lattice of periods $(1, \omega)$. The structure of (3.41) allows one to show that the two-magnon wave function is given by the formula

$$\psi(n_1, n_2) = \frac{e^{i(p_1 n_1 + p_2 n_2)}\sinh[\kappa(n_1 - n_2) + \gamma] + e^{i(p_1 n_2 + p_2 n_1)}\sinh[\kappa(n_1 - n_2) - \gamma]}{\sinh\kappa(n_1 - n_2)} \qquad (3.42)$$

the corresponding energy is

$$\varepsilon^{(2)}(p_1, p_2) = \varepsilon(p_1) + \varepsilon(p_2),$$

where $\varepsilon(p_i)$ are given by (3.40) and the phase γ is connected with pseudomomenta $p_{1,2}$ by the relation

$$\coth\gamma = \frac{1}{2\kappa}\left[\zeta_1\left(\frac{ip_2}{2\kappa}\right) - \zeta_1\left(\frac{ip_1}{2\kappa}\right) + \frac{p_1 - p_2}{\pi}\zeta_1\left(\frac{i\pi}{2\kappa}\right)\right]. \qquad (3.43)$$

This gives, in the limit of $\kappa \to \infty$ ($\omega \to 0$), just the expression for the Bethe phase [4], and the additivity of magnon energies takes place.

Equation (3.43) can be rewritten in the form

$$\coth \gamma = \frac{f(p_1) - f(p_2)}{2\kappa},$$

where

$$f(p) = \frac{p}{\pi} \zeta_1 \left(\frac{i\pi}{2\kappa} \right) - \zeta_1 \left(\frac{ip}{2\kappa} \right). \tag{3.44}$$

It admits also the representation

$$f(p) = i\kappa \cot \frac{p}{2} - \kappa \sum_{n=1}^{\infty} \left[\coth \left(\frac{ip}{2} + \kappa n \right) + \coth \left(\frac{ip}{2} - \kappa n \right) \right].$$

If $p_{1,2}$ are real, the wave function (3.42) describes scattering of magnons. The relatively simple form of (3.43) allows one to investigate the bound states of two magnons in detail [34]. That is, in these states the wave function must vanish as $|n_1 - n_2| \to \infty$. It means that p_1 and p_2 should be complex with $P = p_1 + p_2$ real. The simplest possibility is given by the choice

$$p_1 = \frac{P}{2} + iq, \quad p_2 = \frac{P}{2} - iq,$$

where q is real, and one can always choose $q > 0$ for convenience. Then the vanishing of $\psi(n_1, n_2)$ as $|n_1 - n_2| \to \infty$ is equivalent to the condition

$$\coth \gamma(p_1, p_2) = \frac{f(p_1) - f(p_2)}{2\kappa} = 1. \tag{3.45}$$

The structure of the function (3.44) is crucial for the analysis. It is easy to see that it is odd and double-quasiperiodic,

$$f(p) = -f(-p), \quad f(p + 2\pi) = f(p), \quad f(p + 2i\kappa) = f(p) + 2\kappa. \tag{3.46}$$

Note that one can always choose $q \leq \kappa$ due to (3.46). Equation (3.45) can be rewritten in the more detailed form

$$F_P(q) = 1 - \frac{1}{2\kappa} \left[\frac{2iq}{\pi} \zeta_1 \left(\frac{i\pi}{2\kappa} \right) \right.$$

$$\left. - \zeta_1 \left(\frac{iP}{4a} - \frac{q}{2\kappa} \right) + \zeta_1 \left(\frac{iP}{4a} + \frac{q}{2\kappa} \right) \right] = 0. \tag{3.47}$$

At fixed real P and q, the function (3.47) is real. Moreover, the relations (3.46) imply the following properties of $F_P(q)$,

$$F_P(0) = 1, \quad F_P(q) = -F_P(2\kappa - q), \quad F_P(q) = -F_P(-q) + 2.$$

One can immediately see that $F_P(\kappa) = 0$, but this zero is unphysical: the wave function in this point vanishes identically. The physical solution, if it exists, must lie in the interval $0 < q < \kappa$. Such a non-trivial zero exists if the derivative of $F_P(q)$ is positive at $q = \kappa$,

$$F'_P(\kappa) = -\frac{i}{\pi\kappa}\zeta_1\left(\frac{i\pi}{2\kappa}\right) + \frac{1}{4\kappa^2}\left[\wp_1\left(\frac{iP}{4\kappa} - \frac{1}{2}\right) + \wp_1\left(\frac{iP}{4\kappa} + \frac{1}{2}\right)\right] > 0. \tag{3.48}$$

This inequality indeed takes place for the values of P within the interval $0 < P < P_{cr}$, $0 < P_{cr} < \pi$ [75]. There should be at least one bound state specified by (3.47). At $P > P_{cr}$, the inequality (3.48) does not hold and there are no bound states of this type (type I).

There is, however, another possibility for getting bound states. Since

$$f(p + i\kappa) = \kappa + i\chi(p), \quad f(p - i\kappa) = -\kappa + i\chi(p),$$

one gets only one real equation for real $\tilde{p}_{1,2}$ if one puts $p_1 = \tilde{p}_1 + i\kappa$, $p_2 = \tilde{p}_2 - i\kappa$,

$$\chi(\tilde{p}_1) - \chi(\tilde{p}_2) = 0. \tag{3.49}$$

Noting that

$$\chi(0) = \chi(\pi) = 0,$$

$$\chi'(0) = \frac{a}{2}\sum_{n=-\infty}^{\infty}\frac{1}{\sinh^2 a(n + \frac{1}{2})} > 0,$$

$$\chi'\left(\frac{\pi}{2}\right) = \frac{a}{2}\sum_{n=-\infty}^{\infty}\frac{1}{\sinh^2\left(\frac{i\pi}{4} + a\left(n + \frac{1}{2}\right)\right)} < 0,$$

it is easy to see that there should be some value $\tilde{p}_0$ at which $\chi(\tilde{p})$ has a maximum on the interval $[0, \pi]$ and the corresponding $\tilde{p}'_0 = 2\pi - \tilde{p}_0$ at which $\chi(\tilde{p})$ has a minimum on the interval $[\pi, 2\pi]$. As a matter of fact, $\tilde{p}_0 = \frac{P_{cr}}{2}$. There are no other extrema of $\chi(\tilde{p})$ on the interval $(0, \pi)$. The presence of a maximum means that the equation

$$\chi(\tilde{p}) = \chi_0$$

has two distinct real roots if $0 \leq \chi_0 < \chi\left(\frac{P_{cr}}{2}\right)$, $\frac{P_{cr}}{2} < \tilde{p}_1 \leq \pi$ and $0 \leq \tilde{p}_2 < \frac{P_{cr}}{2}$. These roots serve also as non-trivial solutions to Equation (3.49) and thus give the bound state of type II in which the wave function oscillates and decays exponentially as $|n_1 - n_2| \to \infty$. For $P_{cr} < P \leq \pi$, such a solution always exists. Similar solutions corresponding to $-\chi(\frac{P_{cr}}{2}) < \chi_0 < 0$ can be found with any $-\pi \leq P < -P_{cr}$.

The above treatment is universal with respect to parameter κ in the interval $0 < \kappa < \infty$. In the nearest-neighbor limit $\kappa \to \infty$, the type II states with complex relative pseudomomentum and oscillating wave function disappear ($P_{cr} \to \pi$) and the result coincides with the known one for the Bethe solution.

3.3.2. *Multimagnon scattering*

After solving the two-magnon problem, it is natural to try to find a way to describe scattering of M magnons with $M \geq 3$, i.e. find solution to the difference equation (or lattice Schrödinger equation)

$$\sum_{\beta=1}^{M} \sum_{s \in \mathbf{Z}_{[n]}} V(n_\beta - s)\psi(n_1, \ldots, n_{\beta-1}, s, n_{\beta+1}, \ldots, n_M)$$

$$= -\psi(n_1, \ldots, n_M)\left[\sum_{\beta \neq \gamma}^{M} V(n_\beta - n_\gamma) + J^{-1}\varepsilon_M - M\varepsilon_0\right], \quad (3.50)$$

where $n \in \mathbf{Z}^M$, the notation $\mathbf{Z}_{[n]}$ is used for the variety $\mathbf{Z} - (n_1, \ldots, n_M)$ and $\varepsilon_0 = \sum_{j \neq 0} V(j)$. The exchange interaction $V(j)$ is of hyperbolic form (1.5).

The first attempt to solve (3.50) for $M > 2$ was made in [59] with the use of the trial solution of the Bethe form and taking into account by semi-empirical way the corrections needed due to non-local form of exchange in (3.50). In this paper, the explicit solutions have been found for $M = 3, 4$, but the regular procedure of getting the solution for higher values of M was not proved rigorously. The rigorous treatment of the solutions to (3.50) has been found later [62]. It is based on the analogy of the solution to (3.50) and correspond- ing solution to the quantum Calogero–Moser M-particle system with the same two-body potential and specific value of the coupling constant, determined by $l = 1$. This analogy is already seen in the form of the two-magnon wave function (3.42) and holds for $M = 3, 4$, too. It was the motivation of the paper [62] to use this analogy in detail.

The solution to M-particle system with hyperbolic potential and coupling constant with $l = 1$ is not simple either. The first integral representation for it has been obtained in [60] and a more simple analytic form based on recurrence operator relation was given in [61]. We shall follow [61, 62] for the description of the M-magnon problem on an infinite lattice.

Let us start from the continuum model (3.3) with the interaction (3.5) and $l = 1$. The solution can be written in the form

$$\chi_p^{(M)(x)} = \exp\left(i\sum_{\mu=1}^{M} ip_\mu x_\mu\right) \varphi_p^{(M)}(x), \qquad (3.51)$$

where $\varphi_p^{(M)}(x)$ is periodic in each x_j,

$$\varphi_p^{(M)}(x) = \varphi_p^{(M)}(x_1, \ldots, x_j + i\pi\kappa^{-1}, \ldots, x_M).$$

In [61], the explicit construction of the differential operator which intertwines (3.3) at (3.5) and $l = 1$ with the usual M-dimensional Laplacian has been proposed, and the functions of the type (3.51) have been represented in the form

$$\chi_p^{(M)} = (x) = D_M\exp\left(i\sum_{\mu=1}^{M} p_\mu x_\mu\right), \quad D_M = Q_M^{1,\ldots,M-1} D_{M-1},$$

$$(3.52)$$

where

$$Q_n^{i_1,\dots,i_m}$$

$$= Q_n^{i_1,\dots,i_{m-1}}\left[\frac{\partial}{\partial x_{i_m}} - \frac{\partial}{\partial x_n} - 2\kappa \coth \kappa(x_{i_m} - x_n)\right]$$

$$+ \sum_{s=1}^{m-1} 2\kappa^2 \sinh^{-2}[\kappa(x_{i_s} - x_{i_m})]Q_n^{i_1,\dots,i_{s-1}i_{s+1},\dots,i_{l-1}}, \quad Q_n = 1.$$

$$(3.53)$$

This double recurrence scheme is very cumbersome because of the presence of multiple differentiations, but it allows one to reduce the construction of $\chi_p^{(M)}(x)$ to a much simple problem of solving the set of linear equations. Indeed, it follows from (3.52) and (3.53) that the function $\varphi_p^M(x)$ from (3.13) can be represented in the form

$$\varphi_p^{(M)}(x) = R(\{\coth \kappa(x_j - x_k)\}), \qquad (3.54)$$

where R is a polynomial in the variables $\{\coth \kappa(x_j - x_l)\}$. As it can be seen from the structure of singularities in (3.3), the function $\varphi_p^M(x)$ has a simple pole of the type $[\sinh \kappa(x_j - x_k)]^{-1}$ at each hyperplane $x_j - x_k = 0$. As a consequence of (3.54), all the limits $\varphi_p^{(M)}(x)$ as $x_j \to \pm\infty$ must be finite. Combining these properties with the periodicity of φ, one arrives at the following formula for the eigenfunctions of the Calogero–Moser operator:

$$\chi_p^{(M)}(x) = \exp\left\{\sum_{\mu=1}^{M}[ip_\mu - \kappa(M-1)]x_\mu\right\}$$

$$\times \prod_{\mu>\nu}^{M} \sinh^{-1}\kappa(x_\mu - x_\nu)S_p^{(M)}(y), \qquad (3.55)$$

where $S_p^{(M)}(y)$ is a polynomial in $y_\mu = \exp(2\kappa x_\mu)$ in which the maximal power of each variable cannot exceed $M - 1$. Hence, this polynomial can be represented in the form

$$S_p^{(M)}(y) = \sum_{m \in D^M} d_{m_1,\dots,m_M}(p) \prod_{\mu=1}^{M} y^{m_\mu}, \qquad (3.56)$$

where D^M is the hypercube in $\mathbf{Z}^M$,

$$m \in D^M \leftrightarrow 0 \leq m_\beta \leq M - 1,$$

and $d_m(p)$ is the set of M^M coefficients; it will be shown, however, that most of them vanish. The eigenvalue condition for the function (3.55) can be written in the form

$$\sum_{\beta=1}^{M} \left[2y_\beta \frac{\partial}{\partial y_\beta} \left(y_\beta \frac{\partial}{\partial y_\beta} + i\kappa^{-1} p_\beta - M + 1 \right) \right.$$

$$\left. - i\kappa^{-1} p_\beta (M-1) + (M-1)(2M-1)/3 \right] S_p^{(M)}$$

$$- \sum_{\beta \neq \rho}^{M} \frac{y_\beta + y_\rho}{y_\beta - y_\rho} \left[y_\beta \frac{\partial}{\partial y_\beta} - y_\rho \frac{\partial}{\partial y_\rho} + \frac{i}{2\kappa} (p_\beta - p_\rho) \right] S_p^{(M)} = 0.$$

$$(3.57)$$

It can be satisfied if for each pair (β, ρ) the polynomial

$$\left[y_\beta \frac{\partial}{\partial y_\beta} - y_\rho \frac{\partial}{\partial y_\rho} + \frac{i}{2\kappa} (p_\beta - p_\rho) \right] S_p^{(M)}$$

is divisible by $(y_\beta - y_\rho)$. With the use of (3.56), this condition gives $(M-1)(2M-1)M^M/2$ linear equations for the coefficients $d_m(p)$,

$$\sum_{n \in \mathbf{Z}} d_{m_1, \ldots, m_\beta+n, \ldots, m_\rho-n, \ldots, m_M}(p)$$

$$\times \left[m_\beta - m_\rho + 2n + \frac{i}{2\kappa} (p_\beta - p_\rho) \right] = 0. \qquad (3.58)$$

The sum over n is finite due to restrictions to the indices of $d_m(p)$. Substituting (3.56) gives also the set of equations

$$\sum_{m \in D} \left(\prod_{\mu=1}^{M} y_\mu^{m_\mu} \right) d_m(p) \left\{ \sum_{\beta=1}^{M} \left[2m_\beta^2 + \frac{2i}{\kappa} p_\beta m_\beta \right. \right.$$

$$\left. - \left(2m_\beta + \frac{i}{\kappa} p_\beta - \frac{2M-1}{3} \right) (M-1) \right]$$

$$\left. - \sum_{\beta \neq \rho}^{M} \frac{y_\beta + y_\rho}{y_\beta - y_\rho} \left[m_\beta - m_\rho + \frac{i}{2\kappa} (p_\beta - p_\rho) \right] \right\} = 0. \qquad (3.59)$$

After performing explicit division by $(y_\beta - y_\rho)$ in (3.59), one gets finally the second system of M^M equations. The structure of the set $d_m(p)$ is specified by following propositions (for a sketch of proofs, see [62]).

Proposition 3.1. $S_p^{(M)}(y)$ *is a homogeneous polynomial of the degree* $M(M-1)/2$.

Proposition 3.2. *The set of* $d_m(p)$ *can be chosen as depending on* p *and* κ *only through combinations* $\kappa^{-1}(p_\mu - p_\nu)$.

Proposition 3.3. *Let* $\{P\}$ *be the following set of numbers* $\{m_\mu\}$: $m_\mu = P\mu - 1$, *where* P *is an arbitrary permutation of the permutation group* π_M *and* $1 \leq \mu \leq M$. *The non-vanishing* $d_m(p)$ *with coinciding values of* $\{m_\mu\}$ *are expressed through* $d_{\{P\}}(p)$. *The latter are determined by the system 3.58 up to some normalization constant* d_0,

$$d_{\{P\}}(p) = d_0 \prod_{\mu<\nu}^{M} \left[1 + \frac{i}{2\kappa}(p_{P_\mu^{-1}} - p_{P_\nu^{-1}}) \right]. \tag{3.60}$$

Proposition 3.4. *Let* $(-1)^P$ *be the parity of the permutation* P. *If* $x_{P_{(\mu+1)}} - x_{P_\mu} \to +\infty$, $1 \leq \mu \leq M-1$, *then*

$$\lim \chi_p^{(M)}(x) \exp\left(-i \sum_{\beta=1} p_\beta x_\beta \right) = (-1)^P 2^{\frac{M(M-1)}{2}} d_{\{P^{-1}\}}(p). \tag{3.61}$$

According to Proposition 3.3, the solutions to (3.58) must obey (3.59), and (3.59) has to be considered as a consequence of (3.58). Direct algebraic proof of this fact is still absent.

The problem is now to solve the equations (3.58). It can be done explicitly for $M = 3, 4$ as follows: let $[\mu_1, \ldots, \mu_M]$ be the permutation $(1 \to \mu_1, \ldots, M \to \mu_M)$ and $r_{\mu\nu} = i(2\kappa)^{-1}(p_\mu - p_\nu)$. Then, at $M = 3$ there are six coefficients of the $d_{\{P\}}$ type, which are calculated by the formula (3.60) as follows:

$$d_{012}(p) = d_0(1 + r_{12})(1 + r_{13})(1 + r_{23}),$$
$$d_{102}(p) = d_0(1 + r_{21})(1 + r_{23})(1 + r_{13}),$$
$$d_{210}(p) = d_0(1 + r_{32})(1 + r_{31})(1 + r_{21}),$$
$$d_{021}(p) = d_0(1 + r_{13})(1 + r_{12})(1 + r_{32}),$$
$$d_{120}(p) = d_0(1 + r_{31})(1 + r_{32})(1 + r_{12}),$$
$$d_{201}(p) = d_0(1 + r_{23})(1 + r_{21})(1 + r_{31}).$$

The only non-vanishing coeffcient of another type is determined from (3.58):

$$d_{111}(p) = d_0(6 - r_{12}^2 - r_{13}^2 - r_{23}^2).$$

At $M = 4$, there are 24 coefficients of $d_{\{P\}}$ type and other non-vanishing terms with coinciding values of indices can be arranged in three sets. The first two are given by elements with three coinciding indices and can be found from (3.58) by using known expressions for $d_{\{P\}}$ type,

$$d_{1113}(p) = d_0(1 + r_{14})(1 + r_{24})(1 + r_{34})(6 - r_{12}^2 - r_{13}^2 - r_{23}^2),$$
$$d_{2220}(p) = d_0(1 + r_{41})(1 + r_{42})(1 + r_{43})(6 - r_{12}^2 - r_{13}^2 - r_{23}^2),$$

and other elements of these sets $d_{1131}(p)$, $d_{1311}(p)$, $d_{3111}(p)$ and $d_{2202}(p)$, $d_{2022}(p)$, $d_{0222}(p)$ can be obtained by the permutations [1243], [1342], [2341] of indices in these expressions. The remaining set consists of the coefficients with two pairs of coinciding indices,

$$d_{1122}(p), d_{2211}(p), d_{2112}(p), d_{1221}(p), d_{1212}(p), d_{2121}(p).$$

They may be determined by (3.58) with the use of known coefficients belonging to the first set,

$$d_{1113}(p)(-2 + r_{34}) + d_{1122}(p)r_{34} + d_{1131}(p)(2 + r_{34}) = 0,$$

and others come from the analogous equations arising after the permutations [3412], [3214], [4123], [1324], [4123] of the indices.

These examples show that the solutions to (3.58) are crucial for determining the whole function $\chi_p^{(M)}(x)$. The question is now to see how these findings can be used for spin problem, i.e. the solution to the difference equation (3.50). The motivation is the striking similarity of the wave functions for $M = 2$. Guided by it, I proposed the multimagnon wave functions similar to the functions like (3.51) with the structure (3.55), which are their properly symmetrized combinations,

$$
\psi(n_1, \ldots, n_M) = \prod_{\mu \neq \nu} [\sinh \kappa(n_\mu - n_\nu)]^{-1} \sum_{P \in \pi_M} (-1)^P \exp\left(i \sum_{\lambda=1}^{M} p_{P\lambda} n_\lambda\right)
$$

$$
\times \sum_{m \in D^M} \tilde{d}_{m_1, \ldots, m_n}(p) \exp\left[\kappa \sum_{\lambda=1}^{M} (2m_{P\lambda} - M + 1) n_\lambda\right],
$$

$$(3.62)$$

where $\{\tilde{d}\}$ is the set of unknown coefficients which might be determined from the M-magnon eigenequation if this Ansatz is correct. To verify the hypothesis (3.62), one has to calculate the left-hand side of the equation (3.50) with a wave function of the form (3.62), as follows:

$$
L(\{n\}) = \kappa^2 \sum_{\beta=1}^{M} \sum_{s \in \mathbf{Z}_{[n]}} [\sinh \kappa(n_\beta - s)]^{-2}
$$

$$
\times \psi(n_1, \ldots, n_{\beta-1}, s, \eta_{\beta+1}, \ldots, n_M)
$$

$$
= \sum_{\beta=1}^{M} \sum_{P \in \pi_M} (-1)^P \left[\prod_{\mu > \nu; \mu, \nu \neq \beta} \sinh \kappa(n_\mu - n_\nu)\right]^{-1} (-1)^{(\beta-1)}
$$

$$
\times \sum_{m \in D^M} \tilde{d}_{m_1, \ldots, m_n}(p)
$$

$$
\times \exp\left\{\sum_{\gamma \neq \beta} [i p_{P\gamma} + \kappa(2m_{P\gamma} - M + 1)] n_\gamma\right\}
$$

$$
\times W(p_{P\beta}, m_{P\beta}\{n\}),
$$

$$(3.63)$$

where

$$W(p, m, \{n\}) = \sum_{s \in \mathbf{Z}_{[n]}} \frac{\kappa^2}{\sinh^2 \kappa(s - n_\beta)} \prod_{\lambda \neq \beta}^{M} \sinh^{-1} \kappa(n_\lambda - s)$$

$$\times \exp\{[ip + \kappa(2m - M + 1)]s\}. \tag{3.64}$$

The sum (3.64) converges for all $m \in D^M$ if $p \in \mathbf{C}$ is restricted to $|\Im mp| < 2\kappa$. The explicit calculation of the sum (3.64) is based on the calculation of the function of a complex parameter $x \in \mathbf{C}$,

$$W_q(x) = \sum_{s \in \mathbf{Z}} \frac{\kappa^2 \exp(qs)}{\sinh^2 \kappa(s - n_\beta + x)} \prod_{\lambda \neq \beta}^{M} [\sinh \kappa(n_\lambda - s - x)]^{-1},$$

$$q = ip + \kappa(2m - M + 1).$$

As it follows from the definition, this function is double-quasiperiodic,

$$W_q(x + i\pi\kappa^{-1}) = \exp[i\pi(M - 1)]W_q(x),$$
$$W_q(x + 1) = \exp(-q)W_q(x).$$

Hence, it can be treated on the torus $\mathbf{T}_1 = \mathbf{C}/\mathbf{Z} + i\pi\kappa^{-1}\mathbf{Z}$, and its only singularity on this torus is the double pole at $x = 0$ which arises from the terms with $s = n_1, \ldots, n_M$. The first three terms of its Laurent decomposition can be found directly from the definition,

$$W_q(x) = b_0 x^{-2} + b_1 x^{-1} + b_2 + O(x),$$

$$b_0 = \exp(qn_\beta) \prod_{\lambda \neq \beta}^{M} [\sinh \kappa(n_\lambda - n_\beta)]^{-1},$$

$$b_1 = \kappa \left\{ b_0 \sum_{\gamma \neq \beta}^{M} \coth \kappa(n_\gamma - n_\beta) - \sum_{\rho \neq \beta} \exp(qn_\rho) \right.$$

$$\left. \times \left[\sinh \kappa(n_\beta - n_\rho) \prod_{\lambda \neq \rho}^{M} \sinh \kappa(n_\lambda - n_\rho) \right]^{-1} \right\},$$

$$b_2 = \kappa^2 \left\{ b_0 \left[-\frac{1}{3} + \frac{M-1}{2} + \frac{1}{2} \sum_{\gamma \neq \delta \neq \beta} \coth(n_\gamma - n_\beta) \coth(n_\delta - n_\beta) \right. \right.$$

$$\left. + \sum_{\gamma \neq \beta} \sinh^{-2}(n_\gamma - n_\beta) \right] - \sum_{\rho \neq \beta} \frac{\exp(qn_\rho)}{\sinh \kappa(n_\beta - n_\rho)}$$

$$\times \prod_{\lambda \neq \rho} [\sinh \kappa(n_\lambda - n_\rho)]^{-1}$$

$$\left. \times \left[\coth \kappa(n_\beta - n_\rho) + \sum_{\gamma \neq \rho}^{M} \coth \kappa(n_\gamma - n_\rho) \right] \right\} + W(p, m, \{n\}).$$

The next step consists in constructing the function $U_q(x)$ with the same quasi-periodicity and singularity at $x = 0$ by using the Weierstrass functions $\wp_1(x)$, $\zeta_1(x)$ and $\sigma_1(x)$ defined on the torus $\mathbf{T}_1$,

$$U_q(x) = -A \frac{\sigma_1(x+r)}{\sigma_1(x-r)} \exp(\delta x)$$

$$\times \{\wp_1(x) - \wp_1(r) + \Delta[\zeta_1(x+r) - \zeta(x) - \zeta(2r) + \zeta(r)]\},$$

where A, r, δ and Δ are some constants and the term in braces is chosen as double periodic and having a zero at $x = r$. Hence, the only singularity of $U_q(x)$ on $\mathbf{T}_1$ is double pole at $x = 0$ for all values of r and Δ.

Using the quasiperiodicity of the Weierstrass sigma function, one gets

$$\frac{\sigma_1(x+r+1)}{\sigma_1(x-r+1)} = \exp(2\eta_1 r) \frac{\sigma_1(x+r)}{\sigma_1(x-r)},$$

$$\frac{\sigma_1(x+r+i\pi\kappa^{-1})}{\sigma_1(x-r+i\pi\kappa^{-1})} = \exp(2\eta_2 r) \frac{\sigma_1(x+r)}{\sigma_1(x-r)},$$

where $\eta_1 = 2\zeta_1(1/2)$ and $\eta_2 = 2\zeta_1(i\pi/2\kappa)$. Comparing these expressions with quasiperiodicity of $W_q(x)$, one finds two equations for r and δ,

$$2\eta_1 r + \delta = -q, \quad 2\eta_2 r + i\pi\kappa^{-1}\delta = i\pi(M-1).$$

Their solution can be easily found with the use of the expression for q and Legendre relation $i\pi\kappa^{-1}\eta_1 - \eta_2 = 2\pi i$,

$$r = -\left(\frac{m}{2} + \frac{ip}{4\kappa}\right), \quad \delta = \kappa\left[M - 1 + \frac{4i}{\pi}r\zeta_1\left(\frac{i\pi}{2\kappa}\right)\right].$$

The Laurent decomposition of $U_q(x)$ at $x = 0$ is obtained with the use of standard expansions of the Weierstrass functions,

$$U_q(x) = A[x^{-2} + (2\zeta_1(r) + \delta - \Delta)x^{-1}$$
$$+ \frac{1}{2}(2\zeta_1(r) + \delta - 2\Delta)(2\zeta_1(r) + \delta)$$
$$+ \Delta(2\zeta_1(r) - \zeta_1(2r)) - \wp_1(r)] + O(x).$$

The function $W_q(x) - U_q(x)$ is analytic on $\mathbf{T}_1$ if A and Δ obey the conditions

$$A = b_0, \quad A(2\zeta_1(r) + \delta - \Delta) = b_1.$$

The only analytic function which is double-quasiperiodic on the torus $\mathbf{T}_1$ is zero due to the Liouville theorem. Comparison of third terms in the decompositions of $W_q(x)$ and $U_q(x)$ gives the explicit expression of b_2 in terms of b_0, r, δ and Δ, as follows:

$$b_2 = b_0[1/2(2\zeta_1(r) + \delta - 2\Delta)(2\zeta_r + \delta) + \Delta(2\zeta_1(r)$$
$$- \zeta_1(2r)) - \wp_1(r)].$$

It allows one to find the explicit expression for the sum (3.64) in terms of $p, m, \{n\}$,

$$W(p, m, \{n\}) = \kappa^2 \left\{ -\exp(qn_\beta) \prod_{\lambda \neq \beta}^{M} [\sinh\kappa(n_\lambda - n_\beta)^{-1} \right.$$

$$\times \left[\frac{(M-1)}{2} + \frac{1}{2}\sum_{\gamma \neq \mu \neq \beta}^{M} \coth\kappa(n_\gamma - n_\beta)\right.$$

$$\times \coth\kappa(n_\mu - n_\beta) + \sum_{\gamma \neq \beta}[\sinh\kappa(n_\gamma - n_\beta)]^{-2}$$

$$-\kappa^{-1}\tilde{f}(r)\sum_{\gamma\neq\beta}^{M}\coth\kappa(n_\gamma-n_\beta)+\kappa^{-2}\tilde{\varepsilon}(r)\Bigg]$$

$$+\sum_{\rho\neq\beta}^{M}\frac{\exp(qn_\rho)}{\sinh\kappa(n_\beta-n_\rho)}\prod_{\gamma\neq\rho}[\sinh\kappa(n_\gamma-n_\rho)]^{-1}$$

$$\times\Bigg[\coth\kappa(n_\beta-n_\rho)+\sum_{\gamma\neq\rho}\coth\kappa(n_\gamma-n_\rho)$$

$$-\kappa^{-1}\tilde{f}(r)\Bigg]\Bigg]\Bigg\},\tag{3.65}$$

where

$$\tilde{f}(r)=\zeta_1(2r)+\delta,$$

$$\tilde{\varepsilon}(r)=-\frac{\kappa^2}{3}-\frac{1}{2}\wp_1(2r)+\frac{1}{2}\tilde{f}(r)^2.$$

It is worth noting that $\tilde{f}$ and $\tilde{\varepsilon}$ are some polynomials in m. Indeed, it follows from the definition of r and δ that

$$r=r_p-\frac{m}{2},\quad\delta=\kappa\left[M-1-\frac{2i}{\pi}m\zeta\left(\frac{i\pi}{2\kappa}\right)\right]+\delta_p,$$

where

$$r_p=-\frac{ip}{4\kappa},\quad\delta_p=\frac{p}{\pi}\zeta\left(\frac{i\pi}{2\kappa}\right).$$

By using quasiperiodicity of $\zeta_1(x)$

$$\zeta_1(x+l)=\zeta_1(x)+2l\zeta(1/2),$$

one can represent the above function as

$$\tilde{f}(r)=f(p)-\kappa(2m+1-M),$$

where

$$f(p)=\zeta_1(2r_p)+\delta_p=\frac{p}{\pi}\zeta_1\left(\frac{i\pi}{2\kappa}\right)-\zeta_1\left(\frac{ip}{2\kappa}\right).\tag{3.66}$$

Note that this function just coincides with the function (3.44) used for analysis of two-magnon scattering. The corresponding formula for $\tilde{\varepsilon}$ reads

$$\tilde{\varepsilon}(r) = \varepsilon(p) - \kappa(2m + 1 - M)f(p) + \frac{\kappa^2}{2}(2m + 1 - M)^2, \qquad (3.67)$$

where

$$\varepsilon(p) = -\frac{\kappa^2}{3} - \frac{1}{2}\wp_1(2r_p) + \frac{1}{2}f^2(p).$$

Now, according to (3.65)–(3.67), the left-hand side (3.65) of the eigenequation can be represented as follows:

$$L(\{n\}) = L_1(\{n\}) + L_2(\{n\}) + L_3(\{n\}),$$

where

$$L_1(\{n\}) = \psi(n_1, \ldots, n_M) \left[\sum_{\beta}^{M} \varepsilon(p_\beta) - \sum_{\beta \neq \gamma}^{M} \frac{\kappa^2}{\sinh^2 \kappa(n_\beta - n_\gamma)} \right],$$

$$L_2(\{n\}) = -\kappa^2 \prod_{\mu > \nu}^{M} [\sinh \kappa(n_\mu - n_\nu)]^{-1}$$

$$\times \sum_{P \in \pi_M} (-1)^P \sum_{m \in D^M} \tilde{d}_{m_1, \ldots, m_M}(p)$$

$$\times \sum_{\beta \neq \rho}^{M} \exp \left[\sum_{\gamma \neq \beta, \rho}^{M} [ip_{P_\gamma} + \kappa(2m_{P_\gamma} - M + 1)]n_\gamma \right]$$

$$\times \exp\{[i(p_{P\beta} + p_{P_\rho}) + 2\kappa(m_{P\beta} + m_{P_\rho} - M + 1)]n_\rho\}$$

$$\times [\sinh \kappa(n_\beta - n_\rho)]^{-1} \left[\coth \kappa(n_\beta - n_\rho) \right.$$

$$+ \sum_{\gamma \neq \rho}^{M} \coth \kappa(n_\gamma - n_\rho) - \kappa^{-1}f(p_{P\beta}) + 2m_{P\beta} - M + 1 \right]$$

$$\times \prod_{\gamma \neq \beta, \rho}^{M} \frac{\sinh \kappa(n_\gamma - n_\beta)}{\sinh \kappa(n_\gamma - n_\rho)}, \qquad (3.68)$$

$$L_3(\{n\}) = -\kappa^2 \prod_{\mu \neq \nu}^{M} [\sinh \kappa(n_\mu - n_\nu)]^{-1}$$

$$\times \sum_{P \in \pi_M} (-1)^P \sum_{m \in D^M} \tilde{d}_{m_1,\ldots,m_M}(p)$$

$$\times \exp \left\{ \sum_{\gamma=1}^{M} [ip_{P_\gamma} + \kappa(2m_{P_\gamma} - M + 1)]n_\gamma \right\}$$

$$\times \left\{ \sum_{\beta=1}^{M} \left[\frac{M-1}{2} - \kappa^{-1}(2m_{P\beta} - M + 1) \right. \right.$$

$$\times f(p_{P\beta}) + \frac{(M - 1 - 2m_{P\beta})^2}{2} \right] - \sum_{\beta \neq \gamma} [\kappa^{-1}f(p_{P\beta})$$

$$+ M - 1 - 2m_{P\beta}] \coth \kappa(n_\gamma - n_\beta)$$

$$\left. + \sum_{\beta \neq \gamma \neq \nu} \coth \kappa(n_\gamma - n_\beta) \coth \kappa(n_\nu - n_\beta) \right\}. \tag{3.69}$$

Now, one can see that $L_1(\{n\})$ exactly coincides with the right-hand side of the equation (3.50) if the M-magnon energy is chosen as

$$\varepsilon_M = J \sum_{\beta=1}^{M} [\varepsilon(p_\beta) - \varepsilon_0]$$

$$= J \sum_{\beta=1}^{M} \left[-\frac{1}{2}\wp_1\left(\frac{ip_\beta}{2\kappa}\right) + \frac{1}{2}f^2(p) - \frac{2i\kappa}{\pi}\zeta_1\left(\frac{i\pi}{2a}\right) \right].$$

The problem consists in finding the conditions under which $L_{2,3}(\{n\})$ vanish. Consider at first Equation (3.68) and denote as Q the transposition $\beta \leftrightarrow \rho$ which does not change all other indices from 1 to M. The sum over permutations in (3.68) can be written in the form

$$L_2(\{n\}) = -\kappa^2 \prod_{\mu > \nu}^{M} [\sinh \kappa(n_\mu - n_\nu)]^{-1} \sum_{m \in D^M} \tilde{d}_{m_1,\ldots,m_M}(p)$$

$$\times \sum_{P \in \pi_M} (-1)^P \sum_{\beta \neq \rho}^{M} [F(P) - F(PQ)],$$

where

$$F(P) = \exp\left[\sum_{\gamma \neq \beta, \rho}^{M} (ip_{P_\gamma} + \kappa(2m_{P_\gamma} - M + 1))n_\gamma\right]$$

$$\times \exp\{[i(p_{P\beta} + p_{P_\rho} + 2\kappa(m_{P\beta} + m_{P_\rho} - M + 1)]n_\rho\}$$

$$\times \sinh^{-1}\kappa(n_\beta - n_\rho) \prod_{\gamma \neq \beta, \rho}^{M} \frac{\sinh\kappa(n_\gamma - n_\beta)}{\sinh\kappa(n_\gamma - n_\rho)}$$

$$\times \frac{1}{2}\left[2m_{P\beta} - \kappa^{-1}f(p_\beta) + \coth\kappa(n_\beta - n_\rho)\right.$$

$$\left. + \sum_{\gamma \neq \rho}\coth\kappa(n_\gamma - n_\rho) - M + 1\right].$$

Note that the only difference of $F(PQ)$ and $F(P)$ is in the first two terms in the last brackets. This allows one to rewrite the last formula as

$$L_2(\{n\}) = -\kappa^2\left(\prod_{\mu > \nu}[\sinh\kappa(n_\mu - n_\nu)]^{-1} \sum_{P \in \pi_M} (-1)^P\right.$$

$$\times \sum_{\beta \neq \rho}\exp\left[\sum_{\gamma \neq \beta, \rho}[ip_{P_\gamma} + \kappa(2m_{P_\gamma} - M + 1)]n_\gamma\right]$$

$$\times \sinh^{-1}\kappa(n_\beta - n_\rho)\prod_{\gamma \neq \beta, \rho}\frac{\sinh\kappa(n_\gamma - n_\beta)}{\sinh\kappa(n_\gamma - n_\rho)}$$

$$\times \sum_{\{m_\gamma\} \in D^M, \gamma \neq P\beta, P_\rho} \sum_{s=0}^{2(M-1)}\exp\{[i(p_{P\beta} + p_{P_\rho})$$

$$+ 2\kappa(s - M + 1)]n_\rho\}[M - |s - M + 1|]^{-1}$$

$$\times \sum_{m_{P\beta} + m_{P_\rho} = s}\sum_{n \in \mathbf{Z}}\tilde{d}_{m_1,\dots,m_{P\beta}+n,\dots,m_{P_\rho}-n,\dots,m_M}$$

$$\times \left[m_{P\beta} - m_{P_\rho} - \frac{1}{2\kappa}(f(p_{P\beta}) - f(p_{P_\rho})) + 2n\right]\right).$$

The comparison of the last sum with (3.58) shows that it vanishes if

$$\tilde{d}_{m_1,\dots,m_M}(p) = d_{m_1,\dots,m_M}(if(p)), \qquad (3.70)$$

where $d_{\{m\}}(if(p))$ is an arbitrary solution to the system (3.58) with p_μ replaced by $if(p_\mu)$, $1 \le \mu \le M$.

The only problem is now transformation of $L_3(\{n\})$. Taking into account the formula

$$\sum_{\beta \ne \gamma \ne \nu}^{M} \coth \kappa(n_\gamma - n_\beta) \coth \kappa(n_\nu - n_\beta) = \frac{1}{3}M(M-1)(M-2)$$

and symmetrizing over β, γ in (3.69), one finds

$$L_3(\{n\}) = -\kappa^2 \prod_{\mu > \nu} [\sinh \kappa(n_\mu - n_\nu) \sum_{P \in \pi_M} (-1)^P$$

$$\times \exp\left[\sum_{\gamma=1}^{M} [ip_\gamma - \kappa(M-1)] n_{P-1_\gamma} \right] R(P, \{n\}),$$

where

$$R(P, \{n\}) = \sum_{m \in D^M} \tilde{d}_{m_1,\dots,m_M}(p) \exp\left(2\kappa \sum_{\nu=1}^{M} n_{P-1_\nu} m_\nu \right)$$

$$\times \left\{ \sum_{\beta=1}^{M} \left[\frac{1}{2}(M-1-2m_\beta)^2 + \frac{M^2-1}{6} \right. \right.$$

$$\left. -\kappa^{-1} f(p_\beta)(2m_\beta - M + 1) \right]$$

$$- \sum_{\beta \ne \gamma} [m_\beta - m_\gamma - (2\kappa)^{-1}(f(p_\beta - f(p_\gamma))]$$

$$\left. \times \coth \kappa(n_{P-1_\beta} - n_{P-1_\gamma}) \right\}.$$

Upon introducing the notation $\exp(2\kappa n_{P-1_\gamma}) = y_\gamma$ at fixed P, one finds

$$
R(P, \{n\}) = \sum_{m \in D^M} \tilde{d}_{m_1,\ldots,m_M}(p) \left(\prod_{\gamma=1}^M y_\gamma^{m_\gamma} \right)
$$

$$
\times \left\{ \sum_{\beta=1}^M \left[2m_\beta^2 - 2m_\beta \kappa^{-1} f(p_\beta) \right. \right.
$$

$$
\left. - \left(2m_\beta - \kappa^{-1} f(p_\beta) - \frac{2M-1}{3} \right)(M-1) \right]
$$

$$
\left. - \sum_{\beta \neq \gamma}^M \frac{y_\beta + y_\gamma}{y_\beta - y_\gamma} [m_\beta - m_\gamma - (2\kappa)^{-1}(f(p_\beta) - f(p_\gamma))] \right\}.
$$

Now it is quite easy to see that replacing $\tilde{d}_{m_1,\ldots,m_M}(p) \to d_{m_1,\ldots,m_M}(p)$, $if(p_\mu) \to p_\mu$ in the right-hand side just gives the left-hand side of Equation (3.59) and vanishes for all $y \in \mathbf{R}^M$ if the set $d_{\{m\}}$ solves the Equation (3.58), i.e. the function $\chi_p^{(M)}$ satisfies the Calogero–Moser eigenequation. Hence, both $L_{2,3}(\{n\})$ vanish under the conditions (3.70) and the Ansatz (3.62) satisfies the eigenvalue problem (3.50).

These lengthy calculations lead to the simple receipt: to get a solution to (3.50), one needs to change the p dependence of the periodic part of the solution to the hyperbolic Calogero–Moser quantum problem as $\{p \to if(p)\}$. The asymptotic behavior of the M-magnon wave function $\psi(n_1, \ldots, n_M)$ (3.24) as $\kappa \to \infty$ or $|n_\mu - n_\nu| \to \infty$ can be found with the use of Proposition 3.4. In the former case, one obtains the usual Bethe Ansatz [4,5] as a consequence of (3.61) and the relation

$$
\lim_{\kappa \to \infty} \kappa^{-1}[f(p_1) - f(p_2)] = i \left(\cot \frac{p_1}{2} - \cot \frac{p_2}{2} \right).
$$

The generalized Bethe Ansatz appears at finite κ when the distances between the positions of down spins tend to infinity as

$n_{P(\lambda+1)} - n_{P\lambda} \to +\infty,\ 1 \le \lambda \le M - 1,$

$$\psi(n_1, \ldots, n_M) = \psi_0 \sum_{Q \in \pi_M} (-1)^{QP} \exp\left(i \sum_{\lambda=1}^{M} p_{Q\lambda} n_\lambda\right)$$

$$\times \prod_{\mu < \nu}^{M} \left\{ 1 - \frac{1}{2\kappa}[f(p_{QP\mu}) - f(p_{QP\nu})] \right\}, \qquad (3.71)$$

where $f(p)$ is given by the formula (3.60). The asymptotic form (3.71) will be used in the next section within the asymptotic Bethe Ansatz scheme of calculations of the properties of the antiferromagnetic ground state of the model.

According to (3.71), the multimagnon scattering matrix is factorized as it should be for integrable models. There is a possibility for the existence of multimagnon bound complexes for which some terms in asymptotic expansion (3.71) vanish. Such a situation does not take place for usual quantum Calogero–Moser systems with hyperbolic interactions where the two-body potential is repulsive.

3.4. Periodic Boundary Conditions and Bethe Ansatz Equations

Imposing periodic boundary conditions (with period N) for the spin chains with inverse square hyperbolic interaction leads to the elliptic form of exchange (3.12). These conditions allow one to treat correctly also the important antiferromagnetic case which corresponds to the positive sign of coupling constant J in (3.12).

The spectrum of one-magnon excitations over the ferromagnetic ground state is now discrete and can be calculated via Fourier transform of the elliptic exchange [68]. Throughout this section, the notation $\omega = i\pi/\kappa$ will be used for the second period of the Weierstrass functions. As in the previous section, we shall consider at first the case $M = 2$, which allows more detailed description.

3.4.1. Two-magnon scattering

As in the case of infinite lattice, the problem consists in finding two-magnon wave function defined by

$$|\psi\rangle = \sum_{n_1 \neq n_2} \psi(n_1, n_2) s_{n_1}^- s_{n_2}^- |0\rangle,$$

where $|\psi\rangle$ is an eigenvector of the Hamiltonian and $|0\rangle$ is the "vacuum" vector with all spins up. The corresponding two-particle problem is now the Lamé equation, and the well-known Hermite form of its solution allows one to guess the Ansatz for the wave function in the form

$$\psi(n_1, n_2) = \frac{\exp[i(p_1 n_1 + p_2 n_2)]\sigma_N(n_1 - n_2 + \gamma) + \exp[i(p_1 n_2 + p_2 n_1)]\sigma_N(n_1 + n_2 - \gamma)}{\sigma_N(n_1 - n_2)}.$$

Since ψ should be periodic in each argument, the parameters $p_{1,2}$ are expressed through the phase γ,

$$p_1 N - i\eta_1 \gamma = 2\pi l_1, \quad p_2 N + i\eta_1 \gamma = 2\pi l_2, \tag{3.72}$$

where $\eta_1 = 2\zeta_N(N/2)$, $\eta_2 = 2\zeta_N(\omega/2)$ and $l_1, l_2 \in \mathbf{Z}$.

The solution to the eigenequation is now based on the formula

$$\sum_{k=0}^{N-1} \wp_N(k+z) \frac{\sigma_N(k-l+\gamma+z)}{\sigma_N(k-l+z)} \exp(i\alpha k)$$

$$= -\frac{\sigma_N(l-\gamma)\sigma_1(z+r_{\alpha\gamma})}{\sigma_N(l)\sigma_1(z-r_{\alpha\gamma})}$$

$$\times \exp\left\{\frac{z}{2\pi i}[\zeta_N(N/2)\zeta_1(\omega/2)\gamma + i\zeta_N(\omega/2)\alpha]\right\}$$

$$\times \left\{\wp_1(z) - \wp_1(r_{\alpha\gamma}) + 2(\zeta_1(z+r_{\alpha\gamma})\right.$$

$$- \zeta_1(z) + \zeta_1(r_{\alpha\gamma}) - \zeta_1(2r_{\alpha\gamma}))$$

$$\times \left[\zeta_1(r_{\alpha\gamma}) + \frac{\zeta_N(l-\gamma) - \zeta_N(l)}{2} - \frac{\exp(i\alpha l)\sigma_N(\gamma)\sigma_N(l)}{2\sigma_N(l-\gamma)}\wp_1(l)\right.$$

$$\left.\left. + \frac{1}{4\pi i}(\zeta_N(N/2)\zeta_1(\omega/2)\gamma + i\zeta_N(\omega/2)\alpha)\right]\right\}.$$

where $l \in \mathbf{Z}$ and α and γ are connected by

$$\exp[i\alpha N + 2\gamma \zeta_N(N/2)] = 1, \quad r_{\alpha\gamma} = -(4\pi)^{-1}[\alpha\omega - i\gamma\zeta_N(\omega/2)].$$

The two-magnon energy is given by

$$\varepsilon_2(p_1, p_2, \gamma) = J\{1/4[f(p_1, \gamma) + f(p_2, -\gamma)]^2 \\ + \varepsilon_0(p_1, \gamma) + \varepsilon_0(p_2, -\gamma) + \wp(\gamma)\},$$

where

$$\varepsilon_0(p, \gamma) = \frac{2}{\omega}[\zeta_1(\omega/2) - N\zeta_N(\omega/2)] - \frac{1}{2}\wp_1\left(\frac{i\eta_2\gamma - p\omega}{2\pi}\right),$$

$$f(p, \gamma) = \zeta_1\left(\frac{i\eta_2\gamma - p\omega}{2\pi}\right) + (i\pi)^{-1}[\eta_2\zeta_1(1/2) + ip\zeta_1(\omega/2)],$$

(3.73)

and $p_{1,2}$ and γ are constrained by

$$f(p_1, \gamma) - f(p_2, -\gamma) - 2\zeta_N(\gamma) = 0. \tag{3.74}$$

With the use of (3.72)–(3.74) and direct computation, it is possible to show that

$$S^+ \sum_{n_1 \neq n_2}^{N} \psi(n_1, n_2) s_{n_1}^- s_{n_2}^- |0\rangle = 0,$$

i.e. these states have the total spin $S = S_z = N/2 - 2$.

It is natural to ask how many solutions Equations (3.72), (3.74) have. The completeness of the set of these solutions means that their number should be equal $N(N-3)/2$ since in two-magnon sector there are N solutions with $\psi(n_1, n_2) = \psi_1(n_1) + \psi_1(n_2)$. Is it possible to evaluate the number of solutions to (4.1), (4.3) analytically? The answer is positive [63]. The sketch of the proof is as follows. The constraint (3.74) can be rewritten as

$$F_{l_1, l_2}(\gamma) = \zeta_1\left(\frac{\gamma - l_1\omega}{N}\right) + \zeta_1\left(\frac{\gamma + l_2\omega}{N}\right) + 2\frac{l_1 - l_2}{N}\zeta_1(\omega/2)$$

$$+ \frac{4\gamma}{\omega}[\zeta_N(\omega/2) - N^{-1}\zeta_1(\omega/2)] - 2\zeta_N(\gamma) = 0.$$

At fixed $l_{1,2}$, it is a transcendental equation for γ.

Let now Λ be the manifold which consists of various sets $\{l_{1,2} \in \mathbf{Z}, \gamma \in \mathbf{C}\}$ and call two sets $\{l_1, l_2, \gamma\}, \{l_1', l_2', \gamma'\} \in \Lambda$ equivalent if the corresponding wave functions coincide up to normalization factor. With the use of (3.72) and quasiperiodicity of sigma functions, one finds that the manifold Λ is equivalent to its submanifold Λ_0 defined by the relations

$$0 \leq l_1 \leq N - 1, \quad l_2 = 0, \quad \gamma \in \mathbf{T}_{N,N\omega}.$$

Let $\{\lambda\}$ be a variety of non-equivalent sets within Λ_0. The question now is: how many non-equivalent sets obeying $F_{l_1,0}(\gamma) = 0$ are in Λ_0? To answer it, let us note that the function $F_{l_1,0}(\gamma)$ is double periodic with periods N and $N\omega$ and there is the relation between ζ functions of periods (N, ω) and $(N, N\omega)$,

$$\zeta_N(x) = \zeta(x) + \sum_{j=1}^{N-1} [\zeta(x + j\omega) - \zeta(j\omega)]$$

$$+ \frac{2x}{\omega} [\zeta_N(\omega/2) - \zeta(N\omega/2)],$$

where $\zeta_N(x)$ is the zeta function defined on the torus $\mathbf{T}_{N,N\omega}$. With the use of scaling relation $\zeta_1(N^{-1}x) = N\zeta(x)$, one can rewrite the constraint $F_{l_1,0}(\gamma) = 0$ in the form

$$-2 \sum_{j=0}^{N-1} \zeta(\gamma - j\omega) + N[\zeta(\gamma - l_1\omega) + \zeta(\gamma)]$$

$$+ 2\zeta \left(\frac{N\omega}{2} \right) (l_1 - N + 1) = 0. \tag{3.75}$$

It is easy to see that at $N > 2$ there are N simple poles of the left-hand side of Equation (3.75) located at $\gamma = j\omega$ and this function is elliptic. Then there should be just N roots of Equation (3.75) within $\mathbf{T}_{N,N\omega}$ and, at first sight, $\{\lambda\}$ consists of N^2 elements. However, some sets with different roots of (3.75) may be equivalent. In fact, one can see that if $\gamma_0 \in \mathbf{T}_{N,N\omega}$ is a root, then

$$\gamma_0' = -\gamma_0 + l_1\omega + N \operatorname{sign}(\mathfrak{Re} \gamma_0) + \frac{N\omega}{2} [1 - \operatorname{sign}(l_1|\omega| - \mathfrak{Im}\gamma_0)]$$

is also a root of (3.75). Moreover, all the solutions to the equation $\gamma_0' = \gamma_0$ are the roots of (3.75). The sets of these solutions are different for N, l_1 even or odd. There are four cases. If both N and ω are even, there are only two of these roots, $(N + l_1\omega)/2$ and $(N + l_1\omega + N\omega)/2$. If both N and l_1 are odd, the additional root $l_1\omega/2$ is present. As N is odd and l_1 even, the additional root is $(N + l_1)\omega/2$. And in the case of even N and odd l, one has four such roots since both $l_1\omega/2$ and $(N + l_1)\omega/2$ obey Equation (3.75).

All these explicit roots are combinations of half-integer periods of the torus $\mathbf{T}_{N,N\omega}$. In this case, the wave function can be simplified and it turns out that $\psi(n_1, n_2)$ vanishes identically for all explicit roots listed above.

The number of all non-trivial and non-equivalent sets $\{l_1, 0, \gamma\}$ in the variety $\{\lambda\}$ can be now easily counted. At even N, there are $N/2$ even $\{l_1\}$ with $(N/2) - 1$ non-equivalent roots and $N/2$ odd $\{l_1\}$ with $(N/2) - 2$ ones. At odd N, there are $(N+1)/2$ even $\{l_1\}$ and $(N-1)/2$ odd $\{l_1\}$ with $(N - 3)/2$ non-equivalent roots in both cases. Hence, the total number of elements in the variety $\{\lambda\}$ equals $N(N - 3)/2$ as it should be, and the non-equivalent solutions to (3.75) provide complete description of non-trivial two-magnon states. It would be of interest to investigate, in the limit of large N, the distribution of non-trivial roots within the torus $\mathbf{T}_{N,N\gamma}$.

The two-magnon wave functions, as it was pointed in [63], also have the Hermite form, but the connection between γ and other parameters is not so simple. It will be shown in two subsequent sections that the analogy of particle and spin dynamics still takes place in the more complicated $M \geq 3$ problems.

3.4.2. *Three-magnon scattering*

In this subsection, we consider the problem of finding three-magnon wave functions. One should also start from the three-particle equation of the Lamé type,

$$H_p^{(3)}\varphi = \left\{-\frac{1}{2}\sum_{\alpha=1}^{2}\left(\frac{\partial}{\partial x_\alpha}\right)^2 + 2[\wp(x_1 - x_2)\right.$$

$$\left. + \wp(x_2 - x_3) + \wp(x_3 - x_1)]\right\}\varphi = E\varphi. \tag{3.76}$$

As $H_p^{(3)}$ commutes with each of the operators $S_j^{(\alpha)}$ shifting jth argument of φ to the period ω_α of the Weierstrass function, it is natural to search for the solutions of (3.76) in accordance with the Floquet–Bloch theory, i.e. as quasiperiodic functions in each argument $\{x_j\}$, as follows:

$$\varphi(x_1 + l_1\omega_\alpha, x_2 + l_2\omega_\alpha, x_3 + l_3\omega_\alpha)$$

$$= \exp\left(i\sum_{j=1}^{3} q_j^{(\alpha)} l_j\right) \varphi(x_1, x_2, x_3), \quad l_1, l_2, l_3 \in \mathbf{Z}. \tag{3.77}$$

The only singularity of $\wp(x)$ on the torus T_{ω_1,ω_2} is the second-order pole at $x = 0$. Then it follows from the eigenequation that $\varphi(x_1, x_2, x_3)$ is meromorphic on $(T_{\omega_1,\omega_2})^3$ with simple poles at $x_1 = x_2, x_2 = x_3, x_3 = x_1$, i.e. it can be represented in the form

$$\varphi(x_1, x_2, x_3) = \frac{\Phi(x_1, x_2, x_3)}{\sigma(x_1 - x_2)\sigma(x_2 - x_3)\sigma(x_3 - x_1)}, \tag{3.78}$$

where $\Phi(x_1, x_2, x_3)$ is analytic on $(T_{\omega_1,\omega_2})^3$. One can show that these conditions (3.77) and (3.78), being combined together, determine the structure of Φ up to eight arbitrary parameters. In particular, the constants $q_\alpha^{(j)}$ in (4.6) cannot be chosen independently but are related by

$$\Omega = \sum_{j=1}^{3} [\omega_1 q_j^{(2)} - \omega_2 q_j^{(1)}] \in 2\pi(\mathbf{Z}\omega_1 + \mathbf{Z}\omega_2).$$

So it is always possible to choose $\{q_j^{(\alpha)}\}$ such that Ω vanishes.

Note that the solutions to (3.76) of the type (3.77) contain only four arbitrary parameters (the common normalization factor and three particle quasimomenta). Hence, finding these solutions is equivalent to some purely algebraic problem, i.e. extracting some four-dimensional manifold from $\{\Phi\}$. To formulate it in a constructive way, one needs an appropriate parametrization of $\{\Phi\}$. It can be chosen as

$$\Phi(x_1, x_2, x_3) = A \exp\left[i \sum_{\alpha=1}^{3} k_\alpha x_\alpha\right] \sigma(x_1 - x_2 + \gamma_{12})$$

$$\times \sigma(x_2 - x_3 + \gamma_{23})\sigma(x_3 - x_1 + \gamma_{31})$$

$$\times [B + \zeta(x_1 - x_2 + \gamma_{12}) + \zeta(x_2 - x_3 + \gamma_{23})$$

$$+ \zeta(x_3 - x_1 + \gamma_{31})]. \tag{3.79}$$

Since the poles of $\zeta(x)$ coincide with zeroes of $\sigma(x)$, the functions of the type (3.79) are analytic on $(T_{\omega_1,\omega_2})^3$ as required. The parameters $q_j^{(\alpha)}$ in (3.79) are easily expressed through $\{k\}$ and $\{\gamma\}$ with the use of (3.77),

$$q_j^{(\alpha)} = k_j \omega_\alpha - i\eta_\alpha(\gamma_{jl} + \gamma_{jm}), \tag{3.80}$$

where the auxiliary phases γ_{21}, γ_{32}, γ_{13} are introduced by the relation

$$\gamma_{jm} = -\gamma_{mj},$$

and (jlm) is an arbitrary combination of the numbers from 1 to 3.

Upon substituting (3.79) and (3.80) into (3.76), one arrives at the equation

$$L(x_1, x_2, x_3)$$

$$= A \exp\left[i \sum_{\alpha=1}^{3} k_\alpha x_\alpha\right] \frac{\sigma(y_{12})\sigma(y_{23})\sigma(y_{31})}{\sigma(z_{12})\sigma(z_{23})\sigma(z_{31})}$$

$$\times \{[B + \zeta(y_{12}) + \zeta(y_{23}) + \zeta(y_{31})]$$

$$\times [-\epsilon + i\zeta(z_{12})(k_1 - k_2) + i\zeta(z_{23})(k_2 - k_3) + i\zeta(z_{31})(k_3 - k_1)$$

$$+ 3(\zeta(z_{12})\zeta(z_{23}) + \zeta(z_{12})\zeta(z_{31}) + \zeta(z_{23})\zeta(z_{31})) - (\zeta(y_{12})$$

$$- \zeta(y_{31}))(ik_1 - \zeta(z_{12}) + \zeta(z_{31})) - (\zeta(y_{23}) - \zeta(y_{12}))$$

$$\times (ik_2 - \zeta(z_{23}) + \zeta(z_{12})) - (\zeta(y_{31}) - \zeta(y_{23}))(ik_3 - \zeta(z_{31})$$

$$+ \zeta(z_{23})) - \zeta^2(y_{12}) - \zeta^2(y_{23}) - \zeta^2(y_{31}) + \zeta(y_{12})\zeta(y_{23})$$

$$+ \zeta(y_{12})\zeta(y_{31}) + \zeta(y_{23})\zeta(y_{31}) + \wp(y_{12}) + \wp(y_{23}) + \wp(y_{31})]$$
$$+ [ik_1 - \zeta(z_{12}) + \zeta(z_{31}) + \zeta(y_{12}) - \zeta(y_{31})][\wp(y_{12}) - \wp(y_{31})]$$
$$+ [ik_2 - \zeta(z_{23}) + \zeta(z_{12}) + \zeta(y_{23}) - \zeta(y_{12})][\wp(y_{23}) - \wp(y_{12})]$$
$$+ [ik_3 - \zeta(z_{31}) + \zeta(z_{23}) + \zeta(y_{31}) - \zeta(y_{23})]$$
$$\times [\wp(y_{31}) - \wp(y_{23})] + \wp'(y_{12}) + \wp'(y_{23}) + \wp'(y_{31})\} = 0, \quad (3.81)$$

where

$$\epsilon = E - \frac{1}{2}(k_1^2 + k_2^2 + k_3^2), \quad z_{\alpha\beta} = x_\alpha - x_\beta, \quad y_{\alpha\beta} = x_\alpha - x_\beta + \gamma_{\alpha\beta}.$$

Note that $L(x_1, x_2, x_3)$, like $\varphi(x_1, x_2, x_3)$, is quasiperiodic in each argument. Let us choose one of them (say, x_1) and fix two others. When treated as a function of x_1, L obeys the relations

$$L(x_1 + \omega_j, x_2, x_3) = \exp(iq_1^{(j)})L(x_1, x_2, x_3), \quad j = 1, 2, \qquad (3.82)$$

with $q_1^{(j)}$ given by (3.80). At $\gamma_{12} - \gamma_{31} \not\subset \mathbf{Z}\omega_1 + \mathbf{Z}\omega_2$, one has

$$q_1^{(1)}\omega_2 - q_1^{(2)}\omega_1 \not\subset 2\pi(\mathbf{Z}\omega_1 + \mathbf{Z}\omega_2).$$

The singularities of L on T_{ω_1,ω_2} at fixed x_2, x_3 are second-order poles at $x_1 = x_2$ and $x_1 = x_3$. Near these points one can write the Laurent decompositions of L as

$$\exp\left(-i\sum_{\alpha=1}^{3} k_\alpha x_\alpha\right) L(x_1, x_2, x_3)$$
$$= \lambda_{-2}^{(j)}(z)(x_1 - x_j)^{-2} + \lambda_{-1}^{(j)}(z)(x_1 - x_j)^{-1}$$
$$+ \lambda_0^{(j)}(z, x_1 - x_j), \quad j = 2, 3, \qquad (3.83)$$

where $z = x_2 - x_3$ and $\lambda_0^{(j)}$ are regular when $x_1 \to x_j$. If one proves that all $\lambda_{-2}^{(j)}(z)$, $\lambda_{-1}^{(j)}(z)$ vanish, then L as a function of x_1 will be analytic on T_{ω_1,ω_2}. But, according to the Liouville theorem, the only analytic function on this torus under these conditions is zero. Hence, (3.81) is equivalent to four simpler equations

$$\lambda_{-2}^{(2)}(z) = 0, \quad \lambda_{-2}^{(3)}(z) = 0, \tag{3.84}$$

$$\lambda_{-1}^{(2)}(z) = 0, \quad \lambda_{-1}^{(3)}(z) = 0. \tag{3.85}$$

Let us first consider Equation (3.84). After calculation of the explicit form of leading singularities in (3.83), they can be written as

$$
\begin{aligned}
&[B + \zeta(\gamma_{12}) + \zeta(z + \gamma_{23}) - \zeta(z - \gamma_{31})][i(k_1 - k_2) \\
&\quad + 2\zeta(\gamma_{12}) + \zeta(z - \gamma_{31}) - \zeta(z + \gamma_{23})] \\
&\quad + \wp(z - \gamma_{31}) + \wp(z + \gamma_{23}) - 2\wp(\gamma_{12}) = 0, \tag{3.86} \\
&[B + \zeta(\gamma_{31}) + \zeta(z + \gamma_{23}) - \zeta(z - \gamma_{12})][i(k_3 - k_1) \\
&\quad + 2\zeta(\gamma_{31}) + \zeta(z - \gamma_{12}) - \zeta(z + \gamma_{23})] \\
&\quad + \wp(z - \gamma_{12}) + \wp(z + \gamma_{23}) - 2\wp(\gamma_{31}) = 0. \tag{3.87}
\end{aligned}
$$

The left-hand sides of these equations are double periodic (i.e. elliptic) functions of z with the first-order poles at $z = -\gamma_{23}$, γ_{31} for (3.86) and $z = -\gamma_{23}$, γ_{12} for (3.87). If the pole residues and constant terms in the Laurent decompositions of these functions equal to zero, then, according to the Liouville theorem, they must vanish identically. These conditions can be expressed in the form of four purely algebraic equations for the parameters $B, \{k\}, \{\gamma\}$,

$$B - i(k_1 - k_2) - \zeta(\gamma_{12}) + 2\zeta(\gamma_{31} + \gamma_{23}) = 0,$$

$$B - i(k_3 - k_1) - \zeta(\gamma_{31}) + 2\zeta(\gamma_{12} + \gamma_{23}) = 0,$$

$$
[B + \zeta(\gamma_{12}) + \zeta(\gamma_{31} + \gamma_{23})][i(k_1 - k_2) + 2\zeta(\gamma_{12}) - \zeta(\gamma_{31} + \gamma_{23})]
$$
$$
- 2\wp(\gamma_{12}) - \wp(\gamma_{31} + \gamma_{23}) = 0,
$$

$$
[B + \zeta(\gamma_{31}) + \zeta(\gamma_{12} + \gamma_{23})][i(k_3 - k_1) + 2\zeta(\gamma_{31})
$$
$$
- \zeta(\gamma_{12} + \gamma_{23})] - 2\wp(\gamma_{31}) - \wp(\gamma_{12} + \gamma_{23}) = 0.
$$

Upon eliminating $k_1 - k_2$ and $k_3 - k_1$, the last two equations can be written as

$$[B + \zeta(\gamma_{12}) + \zeta(\gamma_{31} + \gamma_{23})]^2 = 2\wp(\gamma_{12}) + \wp(\gamma_{31} + \gamma_{23}), \qquad (3.88)$$

$$[B + \zeta(\gamma_{31}) + \zeta(\gamma_{12} + \gamma_{23})]^2 = 2\wp(\gamma_{31}) + \wp(\gamma_{12} + \gamma_{23}). \qquad (3.89)$$

Since the difference of (3.88) and (3.89) is linear in B, this parameter can also be expressed easily through the phases $\{\gamma\}$, as follows:

$$\begin{aligned}
B = &-[\zeta(\gamma_{12}) + \zeta(\gamma_{23}) + \zeta(\gamma_{31})] + 3[\wp(\gamma_{12}) - \wp(\gamma_{23})] \\
&\times [\wp(\gamma_{23}) - \wp(\gamma_{31})][\wp(\gamma_{31}) - \wp(\gamma_{12})] \\
&\times \{\wp'(\gamma_{12})[\wp(\gamma_{23}) - \wp(\gamma_{31})] + \wp'(\gamma_{23})[\wp(\gamma_{31}) - \wp(\gamma_{12})] \\
&+ \wp'(\gamma_{31})[\wp(\gamma_{12}) - \wp(\gamma_{23})]\}^{-1}. \qquad (3.90)
\end{aligned}$$

After the substitution of (3.90) into one of Equations (3.88) and (3.89) one gets the nonlinear constraint for γ_{12}, γ_{23}, γ_{31}. At first sight it seems to be enormously cumbersome but can be essentially simplified with the use of addition theorems for zeta functions,

$$\begin{aligned}
\zeta(x) &+ \zeta(y) + \zeta(z) - \zeta(x + y + z) \\
&= 2[\wp(x) - \wp(y)][\wp(y) - \wp(z)][\wp(z) - \wp(x)] \\
&\times \{\wp'(x)[\wp(y) - \wp(z)] + \wp'(y)[\wp(z) - \wp(x)] \\
&+ \wp'(z)[\wp(x) - \wp(y)]\}^{-1}, \\
\wp(x) - \wp(y) &= [\zeta(x) - \zeta(y) - \zeta(x + z) + \zeta(y + z)] \\
&\times [\zeta(x) + \zeta(y) + \zeta(z) - \zeta(x + y + z)],
\end{aligned}$$

which are valid for all $x, y, z \in \mathbf{C}$. The result of simple but tedious calculations is

$$\begin{aligned}
[\zeta(\gamma_{12}) &+ \zeta(\gamma_{23}) + \zeta(\gamma_{31}) - \zeta(\gamma_{12} + \gamma_{23} + \gamma_{31})] \\
&\times \{9\zeta(\gamma_{12} + \gamma_{23} + \gamma_{31}) - 4[\zeta(\gamma_{12} + \gamma_{23}) + \zeta(\gamma_{23} + \gamma_{31}) \\
&+ \zeta(\gamma_{12} + \gamma_{31})] - [\zeta(\gamma_{12}) + \zeta(\gamma_{23}) + \zeta(\gamma_{31})]\} = 0.
\end{aligned}$$

All zeroes of the first factor coincide with the poles of the second one as it follows from the following relation:

$$\zeta(x) + \zeta(y) + \zeta(z) - \zeta(x + y + z) = \frac{\sigma(x + y)\sigma(y + z)\sigma(z + x)}{\sigma(x)\sigma(y)\sigma(z)\sigma(x + y + z)}.$$

Hence, this factor must be omitted and one finally gets the constraint for $\{\gamma\}$ in the form

$$\zeta(\gamma_{12}) + \zeta(\gamma_{23}) + \zeta(\gamma_{31}) + 4[\zeta(\gamma_{12} + \gamma_{23}) + \zeta(\gamma_{23} + \gamma_{31})$$
$$+ \zeta(\gamma_{31} + \gamma_{12})] - 9\zeta(\gamma_{12} + \gamma_{23} + \gamma_{31}) = 0. \tag{3.91}$$

The left-hand side of (3.91) as a function of one of the phases (say, γ_{12}) at fixed values of two others is double periodic and has four simple poles on $\mathbf{C}/\mathbf{Z}\omega_1 + \mathbf{Z}\omega_2$. So there are four roots of Equation (3.91) and γ_{12} can be treated as a four-valued function of γ_{23} and γ_{31} on $(\mathbf{C}/\mathbf{Z}\omega_1 + \mathbf{Z}\omega_2)^2$. The analysis of degenerate cases $\omega_1, \omega_2 \to \infty$ shows that the uniformization problem for the constraint (3.91) is relatively complicated and its solution seems to be very non-trivial.

With the use of the above relations one can express the parameter B and the differences $\{k_\alpha - k_\beta\}$ through the phases $\{\gamma\}$. The result reads

$$B = \frac{1}{2}[\zeta(\gamma_{12}) + \zeta(\gamma_{23}) + \zeta(\gamma_{31}) - 3\zeta(\gamma_{12} + \gamma_{23} + \gamma_{31})],$$

$$i(k_\alpha - k_\beta) = \frac{1}{2}[\zeta(\gamma_{\beta\delta}) - \zeta(\gamma_{\alpha\delta}) - \zeta(\gamma_{\alpha\beta}) + 4\zeta(\gamma_{\beta\delta} - \gamma_{\alpha\delta})$$
$$- 3\zeta(\gamma_{\alpha\beta} + \gamma_{\beta\delta} - \gamma_{\alpha\delta})],$$

where $(\alpha\beta\delta)$ is an arbitrary permutation of (123).

The next step in constructing the solutions to (3.76) consists in solving Equation (3.85). Their explicit form is as follows:

$$\lambda_{-1}^{(2)}(z) = [B + \zeta(\gamma_{12}) + \zeta(z + \gamma_{23}) - \zeta(z - \gamma_{31})]$$
$$\times \{-\epsilon + 3\wp(z) + \wp(z + \gamma_{23})$$
$$- \zeta^2(\gamma_{12}) - \wp(\gamma_{12}) + \zeta(\gamma_{12})[\zeta(z + \gamma_{23}) - \zeta(z - \gamma_{31})]$$
$$- [\zeta(z + \gamma_{23}) - \zeta(z - \gamma_{31})]^2$$

$$- 3[\zeta(z) - \zeta(z - \gamma_{31})][\zeta(z) - \zeta(z + \gamma_{23})] + i(k_1 - k_3)[\zeta(z)$$

$$- \zeta(\gamma_{12}) - \zeta(z - \gamma_{31})] + i(k_2 - k_3)$$

$$\times [\zeta(z) + \zeta(\gamma_{12}) - \zeta(z + \gamma_{23})]\} - \wp'(\gamma_{12})$$

$$+ \wp'(z + \gamma_{23}) + [\wp(z - \gamma_{31}) - \wp(z + \gamma_{23})]$$

$$\times [i(k_3 - k_2) + \zeta(\gamma_{12}) - \zeta(z - \gamma_{31})$$

$$- 2\zeta(z + \gamma_{23}) + 3\zeta(z)] = 0, \tag{3.92}$$

$$\lambda_{-1}^{(3)}(z) = [B + \zeta(\gamma_{31}) + \zeta(z + \gamma_{23}) - \zeta(z - \gamma_{12})]$$

$$\times \{-\epsilon + 3\wp(z) + \wp(z + \gamma_{23})$$

$$- \zeta^2(\gamma_{31}) - \wp(\gamma_{31}) + \zeta(\gamma_{31})[\zeta(z + \gamma_{23}) - \zeta(z - \gamma_{12})]$$

$$- [\zeta(z + \gamma_{23}) - \zeta(z - \gamma_{12})]^2 - 3[\zeta(z) - \zeta(z - \gamma_{12})][\zeta(z)$$

$$- \zeta(z + \gamma_{23})] + i(k_2 - k_1)[\zeta(z) - \zeta(\gamma_{31})$$

$$- \zeta(z - \gamma_{12})] + i(k_2 - k_3)[\zeta(z) + \zeta(\gamma_{31}) - \zeta(z + \gamma_{23})]\}$$

$$- \wp'(\gamma_{31}) + \wp'(z + \gamma_{23}) + [\wp(z - \gamma_{12}) - \wp(z + \gamma_{23})]$$

$$\times [i(k_3 - k_2) + \zeta(\gamma_{31}) - \zeta(z - \gamma_{12})$$

$$- 2\zeta(z + \gamma_{23}) + 3\zeta(z)] = 0. \tag{3.93}$$

It is easy to see that both $\lambda_{-1}^{(2)}(z)$ and $\lambda_{-1}^{(3)}(z)$ are elliptic functions of z with poles at $z = 0$, $-\gamma_{23}$, γ_{31} for (3.92) and at $z = 0$, $-\gamma_{23}$, γ_{12} for (4.22). At arbitrary values of $\{B, \{k\}, \{\gamma\}$ the pole at $z = 0$ is simple and two others are of second order. With the use of (3.86) and (3.87) one can show, however, that $\lambda_{-1}^{(2)}$ and $\lambda_{-1}^{(3)}$ are analytic at $z = 0$ and the remaining poles are simple. Hence, according to the Liouville theorem, $\lambda_{-1}^{(2)}$ and $\lambda_{-1}^{(3)}$ must vanish if their residues at $z = \gamma_{23}$ and values at $z = 0$. In other words, (3.92) and (3.93) are reduced to the following four algebraic equations,

$$\epsilon + i(k_1 - k_3)[-\zeta(\gamma_{31}) + \zeta(\gamma_{12})] + i(k_2 - k_3)[\zeta(\gamma_{31} + \gamma_{23})$$

$$- \zeta(\gamma_{12}) - \zeta(\gamma_{31})] + 2\wp(\gamma_{31} + \gamma_{23}) - 3\wp(\gamma_{31}) + \wp(\gamma_{12})$$

$$+ \zeta^2(\gamma_{12}) + \zeta^2(\gamma_{31} + \gamma_{23}) + 3\zeta^2(\gamma_{31})$$

$$- \zeta(\gamma_{31} + \gamma_{23})[\zeta(\gamma_{12}) + 3\zeta(\gamma_{31})] + [B + \zeta(\gamma_{12}) + \zeta(\gamma_{31} + \gamma_{23})]$$
$$\times [-i(k_1 - k_3) + 3\zeta(\gamma_{31}) - \zeta(\gamma_{12}) - \zeta(\gamma_{31} + \gamma_{23})] = 0,$$

$$[B + \zeta(\gamma_{12}) + \zeta(\gamma_{23}) + \zeta(\gamma_{31})][-\epsilon - 3\wp(\gamma_{31}) - 2\wp(\gamma_{23}) - \wp(\gamma_{12})$$
$$- \zeta^2(\gamma_{12}) - (\zeta(\gamma_{23}) + \zeta(\gamma_{31}))^2 + \zeta(\gamma_{12})(\zeta(\gamma_{23}) + \zeta(\gamma_{31}))$$
$$+ 3\zeta(\gamma_{23})\zeta(\gamma_{31}) + i(k_1 - k_3)(\zeta(\gamma_{31}) - \zeta(\gamma_{12})) + i(k_2 - k_3)(\zeta(\gamma_{12})$$
$$- \zeta(\gamma_{23}))] + [\wp(\gamma_{31}) - \wp(\gamma_{23})][i(k_1 - k_3) + \zeta(\gamma_{12}) + \zeta(\gamma_{23})$$
$$- 2\zeta(\gamma_{31})] - 3\wp'(\gamma_{31}) - 2\wp'(\gamma_{23}) - \wp'(\gamma_{12}) = 0,$$

$$\epsilon + i(k_2 - k_1)[-\zeta(\gamma_{12}) + \zeta(\gamma_{31})] + i(k_2 - k_3)[\zeta(\gamma_{12} + \gamma_{23})$$
$$- \zeta(\gamma_{12}) - \zeta(\gamma_{31})] + 2\wp(\gamma_{12} + \gamma_{23}) - 3\wp(\gamma_{12}) + \wp(\gamma_{31})$$
$$+ \zeta^2(\gamma_{31}) + \zeta^2(\gamma_{12} + \gamma_{23}) + 3\zeta^2(\gamma_{12}) - \zeta(\gamma_{12} + \gamma_{23})$$
$$\times [\zeta(\gamma_{31}) + 3\zeta(\gamma_{12})] + [B + \zeta(\gamma_{31}) + \zeta(\gamma_{12} + \gamma_{23})]$$
$$\times [-i(k_2 - k_1) + 3\zeta(\gamma_{12}) - \zeta(\gamma_{31}) - \zeta(\gamma_{12} + \gamma_{23})] = 0,$$

$$[B + \zeta(\gamma_{12}) + \zeta(\gamma_{23}) + \zeta(\gamma_{31})][-\epsilon - 3\wp(\gamma_{12}) - 2\wp(\gamma_{23}) - \wp(\gamma_{31})$$
$$- \zeta^2(\gamma_{31})] - (\zeta(\gamma_{23}) + \zeta(\gamma_{12}))^2 + \zeta(\gamma_{31})(\zeta(\gamma_{12}) + \zeta(\gamma_{23}))$$
$$+ 3\zeta(\gamma_{23})\zeta(\gamma_{12}) + i(k_2 - k_1)(\zeta(\gamma_{12}) - \zeta(\gamma_{31}))$$
$$+ i(k_2 - k_3)(\zeta(\gamma_{31}) - \zeta(\gamma_{23}))] + [\wp(\gamma_{12}) - \wp(\gamma_{23})]$$
$$\times [i(k_2 - k_1) + \zeta(\gamma_{31}) + \zeta(\gamma_{23}) - 2\zeta(\gamma_{12})] - 3\wp'(\gamma_{12})$$
$$- 2\wp'(\gamma_{23}) - \wp'(\gamma_{31}) = 0.$$

Their essential simplification can be made if the formula

$$\wp'(x) + 2\wp'(y) + 3\wp'(z)$$
$$= 2[\wp(z) - \wp(y)][\zeta(x) + \zeta(y) - \zeta(x + y)]$$
$$\quad + [\zeta(x + y + z) - \zeta(x) - \zeta(y) - \zeta(z)]\{6[\wp(x) + \wp(y) + \wp(z)]$$
$$\quad + 2[\zeta(x) + \zeta(y) - \zeta(x + y)][\zeta(x + y) + \zeta(x + z)$$
$$\quad + \zeta(y + z) - 2\zeta(x) - 2\zeta(y) - 2\zeta(z)]\}$$

is taken into account. Finally, with the use of (3.90) and (3.91) it is possible after long calculations to show that all these equations are mutually equivalent. Any extra constraints on the parameters $\{\gamma\}$ are absent and these equations determine the three-particle energy,

$$
\begin{aligned}
E = \;& \frac{1}{2}(k_1^2 + k_2^2 + k_3^2) - \frac{1}{2}[\zeta^2(\gamma_{12}) + \zeta^2(\gamma_{23}) + \zeta^2(\gamma_{31})] \\
& + \frac{3}{2}\zeta(\gamma_{12} + \gamma_{23} + \gamma_{31})[\zeta(\gamma_{12}) + \zeta(\gamma_{23}) + \zeta(\gamma_{31})] \\
& - 2[\zeta(\gamma_{12})\zeta(\gamma_{23} + \gamma_{31}) + \zeta(\gamma_{23})\zeta(\gamma_{12} + \gamma_{31}) \\
& + \zeta(\gamma_{31})\zeta(\gamma_{12} + \gamma_{23})] + 2[\wp(\gamma_{12}) + \wp(\gamma_{23}) + \wp(\gamma_{31})] \\
& - \frac{9}{4}[\zeta(\gamma_{12} + \gamma_{23} + \gamma_{31}) - \zeta(\gamma_{12}) - \zeta(\gamma_{23}) - \zeta(\gamma_{31})]^2.
\end{aligned}
$$

Further simplification of E can be reached by extracting the energy of the center of mass motion with a total momentum $K = k_1 + k_2 + k_3$ and using expressions for relative quasi-momenta $\{k_\alpha - k_\beta\}$ in combination with addition theorems for the Weierstrass functions,

$$
\begin{aligned}
E = \;& \frac{K^2}{6} + \frac{1}{12}\{27\wp(\gamma_{12} + \gamma_{23} + \gamma_{31}) - 8[\wp(\gamma_{12} + \gamma_{23}) + \wp(\gamma_{23} + \gamma_{31}) \\
& + \wp(\gamma_{12} + \gamma_{31})] - [\wp(\gamma_{12}) + \wp(\gamma_{23}) + \wp(\gamma_{31})]\}.
\end{aligned}
$$

Equations (3.79), (3.90) give a complete solution to the problem of finding the quasiperiodic functions $\varphi(x_1, x_2, x_3)$ obeying (3.76). Note also that one can write $\varphi's$ only in terms of sigma functions,

$$
\begin{aligned}
& \varphi(x_1, x_2, x_3) \\
& = \frac{A\exp(i\sum_{\alpha=1}^{3} k_\alpha x_\alpha)}{\sigma(x_1 - x_2)\sigma(x_2 - x_3)\sigma(x_3 - x_1)} \\
& \quad \times \left\{ \frac{\sigma(x_1 - x_2 + \gamma_{12})\sigma(x_2 - x_3 + \gamma_{23})\sigma(x_3 - x_1 + \gamma_{31})}{2\sigma(\gamma_{12})\sigma(\gamma_{23})\sigma(\gamma_{31})} \right. \\
& \quad \left. + \frac{\sigma(x_1 - x_2 + \gamma_{12} - \Delta)\sigma(x_2 - x_3 + \gamma_{23} - \Delta)\sigma(x_3 - x_1 + \gamma_{31} - \Delta)}{\sigma(\gamma_{12} - \Delta)\sigma(\gamma_{23} - \Delta)\sigma(\gamma_{31} - \Delta)} \right\},
\end{aligned}
$$

where $\Delta = \gamma_{12} + \gamma_{23} + \gamma_{31}$ and the connections between $k's$ and $\gamma's$ are implied. The resemblance of (3.79) to the classical Hermite form of the solution to the Lamè equation is evident.

To start with the three-magnon problem, let us note first that the lattice eigenequation always has solutions of the type

$$\psi_d^{(M)}(n_1 \cdots n_M) = \sum_{\beta=1}^{M} \psi^{(M-1)}(n_1 \cdots n_{\beta-1} n_{\beta+1} \cdots n_M),$$

$$E_M = E_{M-1}, \tag{3.94}$$

where $\psi^{(M-1)}(n_1 \cdots n_{M-1})$ obeys the equation of the type (3.83) with M replaced by $M - 1$. These wave functions describe the M-magnon states which are generated from $(M - 1)$-magnon ones by the action of the component of total spin operator reducing the absolute value of S_z. The remaining M-magnon states corresponding to the lowest possible eigenvalue of the total spin are described by ψs, which cannot be presented in the form (3.94). Now, let us construct three-magnon states of this type in terms of the Ansatz which is similar to the symmetrized three-particle wave function (3.79),

$$\psi^{(3)}(n_1, n_2, n_3)$$

$$= \sum_{P \in \pi_3} \frac{\exp(i \sum_{\alpha=1}^{3} k_\alpha n_{P\alpha})}{\sigma(n_{P1} - n_{P2}) \sigma(n_{P2} - n_{P3}) \sigma(n_{P3} - n_{P1})}$$

$$\times \left[B + \frac{\partial}{\partial \gamma_{12}} + \frac{\partial}{\partial \gamma_{23}} + \frac{\partial}{\partial \gamma_{31}} \right] \sigma(n_{P1} - n_{P2} + \gamma_{12})$$

$$\times \sigma(n_{P2} - n_{P3} + \gamma_{23}) \sigma(n_{P3} - n_{P1} + \gamma_{31}). \tag{3.95}$$

Here, π_3 is the group of all permutations of the numbers from 1 to 3. The Weierstrass functions are defined on the torus $T_{N,\omega}$. Unlike the particle case, ks and γs have to be restricted by the periodicity of ψ in each argument,

$$N k_\alpha - i \eta_1 (\gamma_{\alpha\beta} + \gamma_{\alpha\delta}) = 2\pi l_\alpha, \quad l_\alpha \in \mathbf{Z}, \tag{3.96}$$

where the auxiliary phases are defined as in the particle case.

To calculate the left-hand side of the lattice Schrödinger equation at $M = 3$ with the use of (3.95) and (3.96), one should represent in a closed form the sum over lattice sites of the following type:

$$W(k, \{\gamma\}, \{l\}) = \sum_{s=1; s \neq l_1, l_2}^{N-1} \exp(iks)\wp(s)$$

$$\times \frac{\sigma(s - l_1 + \gamma_1)\sigma(s - l_2 + \gamma_2)}{\sigma(s - l_1)\sigma(s - l_2)}, \quad l_1, l_2 \in \mathbf{Z},$$

(3.97)

where k, γ_1 and γ_2 are chosen so as to ensure the periodicity of the summands in (3.97) with respect to s,

$$kN - i\eta_1(\gamma_1 + \gamma_2) = 0 (\mathrm{mod}\, 2\pi).$$

It can be done by using the technique based on the Liouville theorem. The result reads

$$W(k, \{\gamma\}, \{l\}) = \left[\prod_{\alpha=1}^{2} \frac{\sigma(l_\alpha - \gamma_\alpha)}{\sigma(l_\alpha)} \right] \left\{ \epsilon(k, \gamma_1, \gamma_2) \right.$$

$$+ f(k, \gamma_1, \gamma_2) \sum_{\alpha=1}^{2} [\zeta(l_\alpha) - \zeta(l_\alpha - \gamma_\alpha)]$$

$$- [\zeta(l_1) - \zeta(l_1 - \gamma_1)][\zeta(l_2) - \zeta(l_2 - \gamma_2)]$$

$$+ \frac{1}{2} \sum_{\alpha=1}^{2} [\wp(l_\alpha - \gamma_\alpha) - \wp(l_\alpha)$$

$$\left. - (\zeta(l_\alpha) - \zeta(l_\alpha))^2] \right\} + 2 \sum_{\alpha=1}^{2} \exp(ikl_\alpha)$$

$$\times \left[\prod_{\beta \neq \alpha}^{2} \frac{\sigma(l_\alpha - l_\beta + \gamma_\beta)}{\sigma(l_\alpha - l_\beta)} \right] \sigma(\gamma_\alpha)$$

$$\times \left\{ \wp(l_\alpha)[f(k, \gamma_1, \gamma_2) - \zeta(\gamma_\alpha) \right.$$

$$\left. - \sum_{\beta \neq \alpha} (\zeta(l_\alpha - l_\beta + \gamma_\beta) - \zeta(l_\alpha - l_\beta))] - \wp'(l_\alpha) \right\},$$

where

$$f(k, \gamma_1, \gamma_2) = \zeta_1 \left(-\frac{k\omega}{2\pi} + \frac{i}{\pi} \zeta \left(\frac{\omega}{2} \right) (\gamma_1 + \gamma_2) \right)$$
$$+ (\pi i)^{-1} \left[ik\zeta_1 \left(\frac{\omega}{2} \right) + 2\zeta_1 \left(\frac{1}{2} \right) \zeta \left(\frac{\omega}{2} \right) (\gamma_1 + \gamma_2) \right],$$

$$\epsilon(k, \gamma_1, \gamma_2) = \frac{1}{2} \left[\wp_1 \left(-\frac{k\omega}{2\pi} + \frac{i}{\pi} \zeta \left(\frac{\omega}{2} \right) (\gamma_1 + \gamma_2) \right) - f^2(k, \gamma_1, \gamma_2) \right],$$

and the notation $\wp_1(x)$, $\zeta_1(x)$ is used for the Weierstrass functions defined on the torus $T_{1,\omega}$.

Now, the left-hand side of the lattice Schrödinger equation at $M = 3$ can be divided into two parts,

$$L(n_1, n_2, n_3) = L_1(n_1, n_2, n_3) + L_2(n_1, n_2, n_3),$$

where

$$L_1(n_1, n_2, n_3) = \sum_{P \in \pi_3} \sum_{\alpha=1}^{3} \chi_\alpha(n_{P1}, n_{P2}, n_{P3}),$$

$$L_2(n_1, n_2, n_3) = \sum_{P \in \pi_3} \sum_{\alpha=1}^{3} [\mu_\alpha^{(1)}(n_{P1}, n_{P2}, n_{P3})$$
$$+ \mu_\alpha^{(2)}(n_{P1}, n_{P2}, n_{P3})],$$

and

$$\chi_1(l_1, l_2, l_3)$$

$$= \exp \left(i \sum_{\alpha=1}^{3} k_\alpha l_\alpha \right) \frac{\sigma(l_1 - l_2 + \gamma_{12})\sigma(l_2 - l_3 + \gamma_{23})\sigma(l_3 - l_1 + \gamma_{31})}{\sigma(l_1 - l_2)\sigma(l_2 - l_3)\sigma(l_3 - l_1)}$$

$$\times \left\{ [B + \zeta(l_2 - l_3 + \gamma_{23}) + \zeta(l_1 - l_2 + \gamma_{12}) \right.$$

$$+ \zeta(l_3 - l_1 + \gamma_{31})] \left[\epsilon(k_1, \gamma_{12}, \gamma_{13}) + (f(k_1, \gamma_{12}, \gamma_{13}) \right.$$

$$- \zeta(\gamma_{12}))(\zeta(l_1 - l_2 + \gamma_{12}) - \zeta(l_1 - l_2)) + (f(k_1, \gamma_{12}, \gamma_{13})$$

$$-\zeta(\gamma_{13}))(\zeta(l_1 - l_3 + \gamma_{13}) - \zeta(l_1 - l_3)) - (\zeta(l_1 - l_2 + \gamma_{12})$$
$$-\zeta(l_1 - l_2))(\zeta(l_1 - l_3 + \gamma_{13}) - \zeta(l_1 - l_3)) - \wp(l_1 - l_2)$$
$$\left. -\wp(l_1 - l_3) - \frac{1}{2}(\zeta^2(\gamma_{12}) + \zeta^2(\gamma_{13}) - \wp(\gamma_{12}) - \wp(\gamma_{13})) \right]$$
$$+ [\zeta(l_1 - l_2 + \gamma_{12}) - \zeta(l_1 - l_2)][\wp(\gamma_{12}) - \wp(l_1 - l_3 + \gamma_{13})]$$
$$+ [\zeta(l_1 - l_3 + \gamma_{13}) - \zeta(l_1 - l_3)][\wp(l_1 - l_2 + \gamma_{12})$$
$$- \wp(\gamma_{13})] - \wp(l_1 - l_2 + \gamma_{12})[f(k_1, \gamma_{12}, \gamma_{13}) - \zeta(\gamma_{12})]$$
$$+ \wp(l_1 - l_3 + \gamma_{13})[f(k_1, \gamma_{12}, \gamma_{13}) - \zeta(\gamma_{13})] - \wp(\gamma_{12})\zeta(\gamma_{12})$$
$$\left. + \wp(\gamma_{13})\zeta(\gamma_{13}) + \frac{1}{2}[\wp'(\gamma_{13}) - \wp'(\gamma_{12})] \right\}, \tag{3.98}$$

$$\mu_1^{(1)}(l_1, l_2, l_3)$$

$$= \exp[i(l_2(k_1 + k_2) + k_3 l_3)]\frac{\sigma(l_2 - l_3 + \gamma_{23})\sigma(l_2 - l_3 + \gamma_{13})}{\sigma^2(l_2 - l_3)}$$
$$\times \sigma(\gamma_{12})\{[B + \zeta(l_2 - l_3 + \gamma_{23}) - \zeta(l_2 - l_3 + \gamma_{13}) + \zeta(\gamma_{12})]$$
$$\times [\wp(l_1 - l_2)(f(k_1, \gamma_{12}, \gamma_{13}) - \zeta(\gamma_{12}) - \zeta(l_2 - l_3 + \gamma_{13})$$
$$+ \zeta(l_2 - l_3)) + \wp'(l_1 - l_2)] + \wp(l_1 - l_2)$$
$$\times [\wp(\gamma_{12}) - \wp(l_2 - l_3 + \gamma_{13})]\}, \tag{3.99}$$

$$\mu_2^{(1)}(l_1, l_2, l_3)$$

$$= \exp[i(l_3(k_1 + k_3) + k_2 l_2)]\frac{\sigma(l_2 - l_3 + \gamma_{23})\sigma(l_2 - l_3 - \gamma_{12})}{\sigma^2(l_2 - l_3)}$$
$$\times \sigma(\gamma_{13})\{[B + \zeta(l_2 - l_3 + \gamma_{23}) - \zeta(l_2 - l_3 - \gamma_{12}) - \zeta(\gamma_{13})]$$
$$\times [\wp(l_3 - l_1)(f(k_1, \gamma_{12}, \gamma_{13}) - \zeta(\gamma_{13}) + \zeta(l_2 - l_3 - \gamma_{12})$$
$$- \zeta(l_2 - l_3)) + \wp'(l_1 - l_3)] + \wp(l_3 - l_1)$$
$$\times [-\wp(\gamma_{13}) + \wp(l_2 - l_3 - \gamma_{12})]\}. \tag{3.100}$$

The quantities $\chi_2, \mu_2^{(1)}, \mu_2^{(2)}$ and $\chi_3, \mu_3^{(1)}, \mu_3^{(2)}$ are obtained from (3.98)–(3.100) by cyclic permutations $(123) \to (231)$, $(123) \to (312)$ of the indices in $\{k_\alpha\}, \{l_\alpha\}, \{\gamma_{\alpha\beta}\}$. Note that $L_2(n_1, n_2, n_3)$ can be transformed as follows. If, for example, Q is the transposition $(1 \leftrightarrow 2)$, then

$$\sum_{P \in \pi_3} \mu_2^{(2)}(n_{P1}, n_{P2}, n_{P3}) = \sum_{PQ \in \pi_3} \mu_2^{(2)}(n_{PQ1}, n_{PQ2}, n_{PQ3})$$

$$= \sum_{P \in \pi_3} \mu_2^{(2)}(n_{P2}, n_{P1}, n_{P3}).$$

Hence, the corresponding terms can be combined as

$$\sum_{P \in \pi_3} \{[\mu_1^{(1)}(n_{P1}, n_{P2}, n_{P3}) + \mu_2^{(2)}(n_{P2}, n_{P1}, n_{P3})]$$

$$+ [\mu_1^{(2)}(n_{P1}, n_{P2}, n_{P3}) + \mu_3^{(1)}(n_{P3}, n_{P2}, n_{P1})]$$

$$+ [\mu_2^{(1)}(n_{P1}, n_{P2}, n_{P3}) + \mu_3^{(2)}(n_{P1}, n_{P3}, n_{P2})]\}.$$

Now, one can see with the use of the explicit expressions (3.98)–(3.100) that the terms in all the three braces of this combination vanish under the conditions which are very similar to the conditions (3.84) for vanishing of the second-order poles in the left-hand side of the three-particle equation,

$$[B + \zeta(\gamma_{12}) + \zeta(z + \gamma_{23}) - \zeta(z - \gamma_{31})][-f(k_1, \gamma_{12}, \gamma_{13})$$

$$+ f(k_2, \gamma_{23}, \gamma_{21}) + 2\zeta(\gamma_{12}) + \zeta(z - \gamma_{31}) - \zeta(z + \gamma_{23})]$$

$$- 2\wp(\gamma_{12}) + \wp(z - \gamma_{31}) + \wp(z + \gamma_{23}) = 0, \qquad (3.101)$$

$$[B + \zeta(\gamma_{31}) + \zeta(z + \gamma_{23}) - \zeta(z - \gamma_{12})][-f(k_3, \gamma_{32}, \gamma_{31})$$

$$+ f(k_1, \gamma_{12}, \gamma_{13}) + 2\zeta(\gamma_{31}) + \zeta(z - \gamma_{12}) - \zeta(z + \gamma_{23})]$$

$$- 2\wp(\gamma_{31}) + \wp(z - \gamma_{12}) + \wp(z + \gamma_{23}) = 0,$$

$$z = n_{P2} - n_{P3}. \qquad (3.102)$$

Really, the Equations (3.86) and (3.87) and (3.101) and (3.102) exactly coincide after changing $k_\alpha \to if(k_\alpha, \gamma_{\alpha\beta}, \gamma_{\alpha\delta})$. Hence, one can use all the techniques of the preceding section for determining the relations between the parameters of the three-magnon wave function (3.95).

Since $L_2(n_1, n_2, n_3)$ vanishes if the conditions (3.101) and (3.102) are fulfilled, the final step of construction consists in the investigation of the structure of $L_1(n_1, n_2, n_3)$. It can be done along the lines of the preceding subsection with some minimal changes. One finds that $\psi(n_1, n_2, n_3)$ obeys the lattice Schrödinger equation if B is expressed through the γs, and ks and γs are connected as in the three-particle case after changing $ik_\alpha \to -f(k_\alpha, \gamma_{\alpha\beta}, \gamma_{\alpha\delta})$ and the three-magnon energy is given by

$$
\begin{aligned}
E_3 = {}& \varepsilon(k_1, \gamma_{12}, \gamma_{13}) + \varepsilon(k_2, \gamma_{21}, \gamma_{23}) + \varepsilon(k_3, \gamma_{31}, \gamma_{32}) \\
&+ \frac{1}{6}[f(k_1, \gamma_{12}, \gamma_{13}) + f(k_2, \gamma_{21}, \gamma_{23}) + f(k_3, \gamma_{31}, \gamma_{32})]^2 \\
&- \frac{1}{12}\{27\wp(\gamma_{12} + \gamma_{23} + \gamma_{31}) - 8[\wp(\gamma_{12} + \gamma_{23}) \\
&+ \wp(\gamma_{23} + \gamma_{31}) + \wp(\gamma_{31} + \gamma_{12})] - \wp(\gamma_{12}) - \wp(\gamma_{23}) - \wp(\gamma_{31})\},
\end{aligned}
$$

where

$$
\begin{aligned}
\varepsilon(k, \gamma_1, \gamma_2) = {}& \frac{2}{\omega}\left[\zeta_1\left(\frac{\omega}{2}\right) - N\zeta\left(\frac{\omega}{2}\right)\right] \\
&- \frac{1}{2}\wp_1\left(-\frac{k\omega}{2\pi} + \frac{i}{\pi}\zeta\left(\frac{\omega}{2}\right)(\gamma_1 + \gamma_2)\right).
\end{aligned}
$$

The relations of the type (3.91) can be simplified by using the periodicity conditions (3.101) and (3.102) and the Legendre relation. The result reads

$$
\begin{aligned}
&\zeta_1\left(\frac{\gamma_{\alpha\beta} + \gamma_{\alpha\delta} - l_{\alpha\omega}}{N}\right) - \zeta_1\left(\frac{\gamma_{\beta\alpha} + \gamma_{\beta\delta} - l_{\beta\omega}}{N}\right) + 2\zeta_1\left(\frac{\omega}{2}\right)\frac{l_\alpha - l_\beta}{N} \\
&+ \frac{2}{N\omega}\left[N\zeta\left(\frac{\omega}{2}\right) - \zeta_1\left(\frac{\omega}{2}\right)\right](2\gamma_{\alpha\beta} + \gamma_{\alpha\delta} - \gamma_{\beta\delta}) \\
&+ \frac{1}{2}[\zeta(\gamma_{\alpha\beta}) + \zeta(\gamma_{\alpha\delta}) - \zeta(\gamma_{\beta\delta}) + 4\zeta(\gamma_{\alpha\delta} - \gamma_{\beta\delta}) \\
&+ 3\zeta(\gamma_{\alpha\beta} + \gamma_{\beta\delta} - \gamma_{\alpha\delta})] = 0, \tag{3.103}
\end{aligned}
$$

where $(\alpha\beta\gamma)$ is an arbitrary combination of (123) and $l_1, l_2, l_3 \in \mathbf{Z}$. The corresponding three-magnon wave functions are determined by the solutions of these three transcendental constraints.

Some problems are still open even for the three-magnon case. In particular, it will be of interest to show that the solutions to (3.103) give a complete set of $\frac{1}{6}N(N-1)(N-5)$ three-magnon eigenvectors of the Hamiltonian (3.1) with total spin $S = S_z = \frac{N}{2} - 3$. In the two-magnon case, the answer to the similar question is positive. The selection of the ground state in $M = 3$ sector for antiferromagnetic type of coupling still demands non-trivial algebraic ideas and needs further detailed investigation.

3.4.3. *Multimagnon states*

As in the preceding section, one has to investigate first the solutions to the usual quantum Calogero–Moser problem with coupling constant $l = 1$ in (3.1) and elliptic two-body potential. The general statements on the structure of many-particle wave function have been proved and explicit result for $M = 3$ has been obtained. The problem of arbitrary l and $M = 3$ has been considered in [66] followed by the analytic expression for arbitrary M [80] in the process of solving the elliptic Knizhnik–Zamolodchikov–Bernard equation. Unfortunately, its form turned out to be so complicated that no explicit calculations were possible for multimagnon wave functions. At $M = 3$, the three-magnon wave function has been found explicitly in the previous subsection, but the calculations were very lengthy and it has not been determined how to generalize the method for $M > 3$. The way to the solution of the M-magnon problem which does not refer to explicit form of the solution to M-particle problem was found later [71,81]. Before describing it, it will be of use to formulate basic facts about the wave functions of the continuous M-particle problem for elliptic two-body interaction [65].

Since $\wp(x)$ is double periodic, it is easy to see that the corresponding M-particle Hamiltonian (3.1) commutes with $2M$ shift operators $Q_{\alpha j} = \exp(\omega_\alpha \partial/\partial x_j)$, where $\omega_{1,2}$ are two periods of $\wp(x)$.

Let $\chi^{(p)}(x_1, \ldots, x_M)$ be their common eigenvector,

$$\chi^{(p)}\left(x_1 + \sum_{\alpha=1}^{2} l_1^{(\alpha)}\omega_\alpha, \ldots, x_M + \sum_{\alpha=1}^{2} l_M^{(\alpha)}\omega_\alpha\right)$$

$$= \exp\left(i \sum_{j=1}^{M}\sum_{\alpha=1}^{2} p_\alpha^{(j)} l_j^{(\alpha)}\right)\chi^{(p)}(x_1, \ldots, x_M),$$

where $l_j^{(\alpha)} \in \mathbf{Z}$. Hence, $\chi^{(p)}(x)$ can be treated on the M-dimensional torus $\mathbf{T}_M = \mathbf{C}/\mathbf{Z}\omega_1 + \mathbf{Z}\omega_2$ with quasiperiodic boundary conditions. The structure of singularities of the Hamiltonian (3.1) on this torus shows that $\chi^{(p)}$ is analytic except for all hypersurfaces L_{jk} defined by the equalities $x_j = x_k, 1 \le j < k \le M$. On each L_{jk}, $\chi^{(q)}$ has a simple pole. Let Ψ_M be a class of functions with these properties.

Proposition 4.1. *The class Ψ_M is a functional manifold of dimension $2M - 1 + M^{M-2}$. The parameters $\{p_\alpha^{(j)}\}$ are not independent but restricted by the linear relation $\sum_{j=1}^{M}(p_1^{(j)}\omega_2 - p_2^{(j)}\omega_1) = \mathbf{Z}\omega_1 + \mathbf{Z}\omega_2$. The manifold Ψ_M can be described as a union of the $(2M - 1)$-parametric family of linear spaces with dimensions M^{M-2} with the basic vectors parametrized by $\{p_\alpha^{(j)}\}$.*

Proposition 4.2. *The co-ordinate system on Ψ_M can be chosen in such a way that all its elements are expressed through the Riemann theta functions of genus 1 or usual Weierstrass sigma functions.*

The sketch of the proofs can be found in [65]. The explicit expressions for $\chi^{(p)}$ can be also found in [65] for $M = 3$ and in [80] for arbitrary M. The amazing fact is that the treatment of M-magnon problem can be done *without* use of these explicit expressions.

Let us choose the exchange in the form

$$h(j) = J\left(\frac{\omega}{\pi}\sin\frac{\pi}{\omega}\right)^2\left[\wp_N(j) + \frac{2}{\omega}\zeta_N\left(\frac{\omega}{2}\right)\right],$$

so as to reproduce correctly the inverse square hyperbolic form of the Section 3.3 in the thermodynamic limit $N \to \infty$. The second period of the Weierstrass function $\wp_N$ is $\omega = i\pi/\kappa$. The eigenproblem is decomposed into the problems with M down spins due to rotation invariance and the eigenvectors $|\psi^{(M)}\rangle$ are given by

$$|\psi^{(M)}\rangle = \sum_{n_1 \ldots n_M}^{N} \psi_M(n_1 \ldots n_M) \prod_{\beta=1}^{M} s_{n_\beta}^{-} |0\rangle,$$

where $|0\rangle = |\uparrow\uparrow \cdots \uparrow\rangle$ is the ferromagnetic ground state with all spins up and the summation is taken over all combinations of integers $\{n\} \leq N$ such that $\prod_{\mu<\nu}^{M}(n_\mu - n_\nu) \neq 0$. The function ψ_M is completely symmetric in its arguments and obeys the lattice Schrödinger equation

$$\sum_{s \neq n_1, \ldots, n_M} \sum_{\beta=1}^{M} \wp_N(n_\beta - s)\psi_M(n_1, \ldots, n_{\beta-1}, s, n_{\beta+1}, \ldots, n_M)$$

$$+ \left[\sum_{\beta \neq \gamma}^{M} \wp_N(n_\beta - n_\gamma) - \mathcal{E}_M\right] \psi_M(n_1, \ldots, n_M) = 0 \qquad (3.104)$$

and the eigenvalues of the Hamiltonian are given by

$$\mathcal{E}_M = J\left(\frac{\omega}{\pi}\sin\frac{\pi}{\omega}\right)^2 \left\{\mathcal{E}_M + \frac{2}{\omega}\left[\frac{2M(2M-1)-N}{4}\zeta_N\left(\frac{\omega}{2}\right)\right.\right.$$

$$\left.\left. - M\zeta_1\left(\frac{\omega}{2}\right)\right]\right\}.$$

Let $\chi_M^{(p)}$ be the special solution to the continuum quantum many-particle problem

$$\left[-\frac{1}{2}\sum_{\beta=1}^{M}\frac{\partial^2}{\partial x_\beta^2} + \sum_{\beta \neq \lambda}^{M} \wp_N(x_\beta - x_\lambda) - \mathcal{E}_M(p)\right] \chi_M^{(p)}(x_1, \ldots, x_M) = 0,$$

which is specified up to some normalization factor by particle pseudomomenta $(p_1, \ldots, p_M)$ and obeys the quasi-periodicity conditions

$$\chi_M^{(p)}(x_1, \ldots, x_\beta + N, \ldots, x_M)$$

$$= \exp(ip_\beta N)\chi_M^{(p)}(x_1, \ldots, x_M), \quad 1 \le \beta \le M,$$

$$\chi_M^{(p)}(x_1, \ldots, x_\beta + \omega, \ldots, x_M)$$

$$= \exp(2\pi i q_\beta(p) + ip_\beta \omega)\chi_M^{(p)}(x_1, \ldots, x_M), \quad 0 \le \Re\mathfrak{e}(q_\beta) < 1.$$

$$(3.105)$$

As will be seen later, the set $\{q_\beta(p)\}$ is completely determined by $\{p\}$.

The connection of $\chi_M^{(p)}$ with multimagnon wave function is given by the Ansatz

$$\psi_M(n_1, \ldots, n_M) = \sum_{P \in \pi_M} \varphi_M^{(p)}(n_{P1}, \ldots, n_{PM}),$$

$$\varphi_M^{(p)}(n_1, \ldots, n_M) = \exp\left(-i\sum_{\nu=1}^{M} \tilde{p}_\nu n_\nu\right) \chi_M^{(p)}(n_1, \ldots, n_M), \quad (3.106)$$

where

$$\tilde{p}_\nu = p_\nu - 2\pi N^{-1} l_\nu, \quad l_\nu \in \mathbf{Z}.$$

The last condition is just the condition of periodicity of ψ_M. The problem now consists in calculation of the left-hand side of the lattice Schrödinger equation (3.104), but before doing this let us mention that $\chi_M(p)$ has the singularities in the form of simple poles and can be presented in the form

$$\chi_M^{(p)} = \frac{F^{(p)}(x_1, \ldots, x_M)}{G(x_1, \ldots, x_M)}, \quad G(x_1, \ldots, x_M) = \prod_{\alpha < \beta}^{M} \sigma_N(x_\alpha - x_\beta),$$

$$(3.107)$$

where $\sigma_N(x)$ is the Weierstrass sigma function on the torus T_N. By definition, the only simple zero of $\sigma_N(x)$ on T_N is located at $x = 0$. Thus, $[G(x_1, \ldots, x_M)]^{-1}$ absorbs all the singularities of $\chi_M^{(p)}$ on the hypersurfaces $x_\alpha = x_\beta$. The numerator $F^{(p)}$ in (3.107) should be analytic on $(T_N)^M$. It obeys the equation

$$\sum_{\alpha=1}^{M} \frac{\partial^2 F^{(p)}}{\partial x_\alpha^2} + \left[2E_M(p) - \frac{M}{2} \sum_{\alpha \neq \beta}^{M} (\wp_N(x_\alpha - x_\beta) - \zeta_N^2(x_\alpha - x_\beta)) \right] F^{(p)}$$

$$= \sum_{\alpha \neq \beta} \zeta_N(x_\alpha - x_\beta) \left(\frac{\partial F^{(p)}}{\partial x_\alpha} - \frac{\partial F^{(p)}}{\partial x_\beta} \right).$$

The left-hand side of this equation is regular as $x_\mu \to x_\nu$. Hence, $F^{(p)}$ must obey the condition

$$\left(\frac{\partial}{\partial x_\mu} - \frac{\partial}{\partial x_\nu} \right) F^{(p)}(x_1, \ldots, x_M)|_{x_\mu = x_\nu} = 0 \qquad (3.108)$$

for any pair (μ, ν). Let us now show that the properties (3.105), (3.107)–(3.108) allow to validate the Ansatz (3.106) for ψ_M. Substitution of (3.106) to (3.104) yields

$$\sum_{p \in \pi_M} \left\{ \sum_{\beta=1}^{M} S_\beta(n_{P1}, \ldots, n_{PM}) + \left[\sum_{\beta \neq \gamma}^{M} \wp_N(n_{P\beta} - n_{P\gamma}) - \varepsilon_M \right] \right.$$

$$\left. \times \varphi_M^{(p)}(n_{P1}, \ldots, n_{PM}) \right\} = 0.$$

where

$$S_\beta(n_{P1}, \ldots, n_{PM}$$

$$= \sum_{s \neq n_{P1}, \ldots, n_{PM}}^{N} \wp_N(n_{P\beta} - s) \hat{Q}_\beta^{(s)} \varphi_M^{(p)}(n_{P1}, \ldots, n_{PM}) \qquad (3.109)$$

and the operator $\hat{Q}_\beta^{(s)}$ replaces βth argument of the function of M variables to s.

The calculation of the sum (3.109) is based on introducing the function of complex variable x

$$W_P^{(\beta)}(x) = \sum_{s=1}^{N} \wp_N(n_{P\beta} - s - x)\hat{Q}_\beta^{(s+x)}\varphi_M^{(p)}(n_{P1}, \ldots, n_{PM}).$$

As a consequence of (3.105), it obeys the relations

$$W_P^{(\beta)}(x+1) = W_P^{(\beta)}(x), \quad W_P^{(\beta)}(x+\omega) = \exp(2\pi i \tilde{q}_\beta(p))W_P^{(\beta)}(x),$$
$$(3.110)$$

where

$$\tilde{q}_\beta(p) = q_\beta(p) + \frac{l_\beta}{N}\omega.$$

The only singularity of $W_P^{(\beta)}$ on the torus $T_1 = \mathbf{C}/\mathbf{Z} + \mathbf{Z}\omega$ is located at the point $x = 0$. It arises from the terms in (3.109) with $s = n_{P1}, \ldots, n_{PM}$. Hence, the Laurent decomposition of $W_P^{(\beta)}$ near $x = 0$ has the form

$$W_P^{(\beta)}(x) = w_{-2}x^{-2} + w_{-1}x^{-1} + w_0 + O(x). \qquad (3.111)$$

With the use of (3.107), one can find the explicit expressions for w_{-i} in the form

$$w_{-2} = \varphi_M^{(p)}(n_{P1}, \ldots, n_{PM})$$

$$w_{-1} = \frac{\partial}{\partial n_{P\beta}}\varphi_M^{(p)}(n_{P1}, \ldots, n_{PM})$$

$$+ (-1)^P G^{-1}(n_1, \ldots, n_M) \sum_{\lambda \neq \beta} T_{\beta\lambda}(n_{P1}, \ldots, n_{PM})\hat{Q}_\beta^{(n_{p\lambda})}$$

$$\times \exp\left(-i\sum_{v=1}^{M} \tilde{p}_v n_{Pv}\right) F^{(p)}(n_{P1}, \ldots, n_{PM})$$

$$w_0 = S_\beta(n_{P1}, \ldots, n_{PM}) + \frac{1}{2}\frac{\partial^2}{\partial n_{P\beta}^2}\varphi_M^{(p)}(n_{P1}, \ldots, n_{PM})$$

$$+ (-1)^P G^{-1}(n_1, \ldots, n_M)$$

$$\times \sum_{\lambda \neq \beta} T_{\beta\lambda}(n_{P1}, \ldots, n_{PM})\left[U_{\beta\lambda}(n_{P1}, \ldots, n_{PM})\hat{Q}_\beta^{(n_{P\lambda})}\right.$$

$$\left. + \wp_N(n_{P\beta} - n_{P\lambda})\partial\hat{Q}_\beta^{(n_{P\lambda})}\right]$$

$$\times \exp\left(-i\sum_{v=1}^{M}\tilde{p}_v n_{Pv}\right)F^{(p)}(n_{P1}, \ldots, n_{PM}),$$

where

$$T_{\beta\lambda}(n_{P1}, \ldots, n_{PM}) = \sigma_N(n_{P\lambda} - n_{P\beta})\prod_{\rho \neq \beta, \lambda}^{M}\frac{\sigma_N(n_{P\rho} - n_{P\beta})}{\sigma_N(n_{P\rho} - n_{P\lambda})},$$

$$U_{\beta\lambda}(n_{P1}, \ldots, n_{PM}) = \wp_N'(n_{P\lambda} - n_{P\beta}) - \wp_N(n_{P\beta} - n_{P\lambda})$$

$$\times \sum_{\rho \neq \beta, \lambda}\zeta_N(n_{P\rho} - n_{P\lambda}),$$

$(-1)^P$ is the parity of the permutation P, and the action of the operator $\partial\hat{Q}_\beta^{(n_{P\lambda})}$ on the function Y of M variables is defined as

$$\partial Q_\beta^{(n_{P\lambda})}Y(z_1, \ldots, z_M) = \frac{\partial}{\partial Z_\beta}Y(z_1, \ldots, z_M)|_{Z_\beta = nP\lambda}.$$

Note now that the expression for the function $W_P^{(\beta)}(x)$ obeying the relations (3.110) and (3.111) can be written analytically without any further freedom as follows:

$$W_P^{(\beta)}(x) = \exp(a_\beta x)\frac{\sigma_1(r_\beta + x)}{\sigma_1(r_\beta - x)}\{w_{-2}(\wp_1(x) - \wp_1(r_\beta))$$

$$+ (w_{-2}(a_\beta + 2\zeta_1(r_\beta)) - w_{-1})$$

$$\times [\zeta_1(x - r_\beta) - \zeta_1(x) + \zeta_1(r_\beta) - \zeta_1(2r_\beta)]\}.$$

The parameters a_β, r_β are chosen so as to satisfy the conditions (3.111) as follows:

$$a_\beta = 2\tilde{q}_\beta(p)\zeta_1(1/2), \quad r_\beta = -\frac{1}{2}\tilde{q}_\beta(p).$$

By expanding the above form of $W_P^{(\beta)}$ in powers of x, one can find w_0 in terms of w_{-2}, w_{-1}, q_β and obtain the explicit expression for $S_\beta(n_{P1}, \ldots, n_{PM})$. After long but straight-forward calculations, Equation (3.104) can be recast in the form

$$\sum_{P \in \pi_M} \left[-\frac{1}{2} \sum_{\beta=1}^{M} \left(\frac{\partial}{\partial n_{P\beta}} - f_\beta(p) \right)^2 + \sum_{\beta \neq \gamma}^{M} \wp_N(n_{P\beta} - n_{P\gamma}) - \mathcal{E}_M \right.$$

$$\left. + \sum_{\beta=1}^{M} \varepsilon_\beta(p) \right] \varphi^{(p)}(n_{P1}, \ldots, n_{PM})$$

$$= \frac{1}{2} G^{-1}(n_1, \ldots, n_M) \sum_{P \in \pi_M} (-1)^P \sum_{\beta \neq \lambda} [Z_{\beta\lambda}(n_{P1}, \ldots, n_{PM})$$

$$+ Z_{\lambda\beta}(n_{P1}, \ldots, n_{PM})], \tag{3.112}$$

where

$$f_\beta(p) = 2\tilde{q}_\beta(p)\zeta_1(1/2) - \zeta_1(\tilde{q}_\beta(p)),$$

$$\varepsilon_\beta(p) = \frac{1}{2}\wp_1(\tilde{q}_\beta(p)),$$

and $Z_{\beta\lambda}(n_{P1}, \ldots, n_{PM})$ is defined by the relation

$$Z_{\beta\lambda}(n_{P1}, \ldots, n_{PM}) = T_{\beta\lambda}(n_{P1}, \ldots, n_{PM})[U_{\beta\lambda}(n_{P1}, \ldots, n_{PM})\tilde{Q}_\beta^{(n_{P\lambda})}$$

$$+ \wp_N(n_{P\lambda} - n_{P\beta}) \times (\partial\hat{Q}_\beta^{(n_{P\lambda})} - f_\beta(p)\hat{Q}_\beta^{(n_{P\lambda})})]$$

$$\times \exp\left(-i \sum_{v=1}^{M} \tilde{p}_v n p_v \right) F^{(p)}(n_{P1}, \ldots, n_{PM}).$$

One observes with the use of the definition (3.106) of $\varphi^{(p)}$ that each term of the left-hand side of (3.112) has the same structure as the left-hand side of the many-particle Schrödinger equation and vanishes if $\mathcal{E}_M$ and $f_\beta(p)$ are chosen as

$$f_\beta(p) = -i\tilde{p}_\beta, \quad \beta = 1, \ldots, M, \tag{3.113}$$

$$\mathcal{E}_M = \mathcal{E}_M(p) + \sum_{\beta=1}^{M} \varepsilon_\beta(p). \tag{3.114}$$

It remains to prove that the right-hand side of (3.112) also vanishes. It can be done by using the observation that the sum over permutations in it can be simply recast in the form

$$\sum_{P \in \pi_M} (-1)^P \sum_{\beta \neq \lambda} [Z_{\beta\lambda}(n_{P1}, \ldots, n_{PM}) - Z_{\lambda\beta}(n_{PR1}, \ldots, n_{PRM})],$$

where R is the transposition $(\beta \leftrightarrow \lambda)$ which leaves other numbers from 1 to M unchanged. Taking into account the definition of Z, one finds

$$Z_{\beta\lambda}(n_{P1}, \ldots, n_{PM}) - Z_{\lambda\beta}(n_{PR1}, \ldots, n_{PRM})$$
$$= T_{\beta\lambda}(n_{P1}, \ldots, n_{PM})\wp_N(n_{P\lambda} - n_{P\beta})$$
$$\times \exp\left[-i\left((\tilde{p}_\beta + \tilde{p}_\lambda)n_{P\lambda} + \sum_{\rho \neq \beta, \lambda}^{M} \tilde{p}_\rho n_{P\rho}\right)\right]$$
$$\times \left(\frac{\partial}{\partial n_{P\beta}} - \frac{\partial}{\partial n_{P\lambda}}\right) F^{(p)}(n_{P1}, \ldots, n_{PM})|_{n_{P\beta}=n_{P\lambda}}.$$

Now it is clearly seen that the last factor vanishes due to the condition (3.108).

The relations (3.113) and (3.114) for the spectrum are still not complete since the dependence of $\{q\}$ on $\{p\}$ is not known on this stage. This completion can be done only by further analysis of the properties of $\chi_M^{(p)}$ solving the M-particle Schrödinger equation. In [39] the explicit form of $\chi_M^{(p)}(x)$ has been found in the process of solving the Knizhnik–Zamolodchikov–Bernard equations. In suitable notations, it reads

$$\chi_M^{(p)}(x) \sim \exp\left(i\sum_{\beta=1}^{M} p_\beta x_\beta\right) \sum_{s\in\pi_m} l(s)$$

$$\times \prod_{j=1}^{m} \tilde{\sigma}_{\sum\limits_{k=1}^{j}(x_c(s(k))-x_c(s(k))+1)}(t_{s(j)} - t_{s(j+1)}), \quad (3.115)$$

where $m = M(M-1)/2, c$ is non-decreasing function $c :$ $\{1,\ldots,m\} \to \{1,\ldots,M-1\}$ such that $|c^{-1}\{j\}| = M-j, l(s)$ is an integer which is defined for the permutation s by the relation $x_{c(s(1))+1}\partial/\partial x_{c(s(1))},\ldots,x_{c(s(m))+1}\partial/\partial x_{c(s(m))}x_1^M = l(s)(x_1,\ldots,x_M), \{t\}$ is a set of m complex parameters obeying m relations [80]

$$\sum_{l:|c(l)-c(j)|=1} \rho(t_j - t_l) - 2 \sum_{l:l\neq j,c(l)-c(j)} \rho(t_j - t_l) + M\delta_{cj,1}\rho(t_j)$$

$$= i(p_{c(j)} - p_{c(j)+1}), \quad (3.116)$$

$$\rho(t) = \zeta_N(t) - \frac{2}{N}\zeta_N(N/2)t,$$

and

$$\tilde{\sigma}_w(t) = \exp((2/N)\zeta_N(N/2)wt)\frac{\sigma_N(w-t)}{\sigma_N(w)\sigma_N(t)}.$$

The main advantage of the explicit form of χ function is that it allows to find the second set of relations between the Bloch factors $\{p\}, \{q\}$. It is easy to see that $\{p\}s$ in the definitions (3.105) and (3.115) are the same. The problem consists in calculation of $\{q\}$. To do this, it is not necessary to analyze each term in the sum over permutations in (3.115) since all of them must have the same Bloch factors. It is convenient to choose the term which corresponds to the permutation

$$s_0 : s_0(j) = m+1-j, \quad j = 1,\ldots,m.$$

After some algebra, one finds that this permutation gives non-trivial contribution to (3.115) with $l(s_0) = M!(M-1)!\cdots 2!$. Moreover, with the use of the explicit form of the color function, one finds

$$c(s_0(l)) = M - q \text{ if } q(q-1)/2 + 1 \leq l \leq q(q+1)/2.$$

Now, the problem of calculation of the second Bloch factors reduces to some long and tedious but in fact simple calculations of the product of factors which various $\tilde{\sigma}$ functions acquire under changing arguments of χ function to the quasiperiod ω. The final result is surprisingly simple,

$$
q_\beta(p) = N^{-1} \left(\sum_{l:c(l)=\beta} t_l - \sum_{l:c(l)=\beta-1} t_l \right), \quad 1 < \beta < M - 1, \quad (3.117)
$$

with the first and second term being omitted for $\beta = M$ and $\beta = 1$.

Equation (3.117), together with (3.113) and (3.116), form a closed set for finding Bloch factors $\{p\}, \{q\}$ at given integers $\{l_\beta\} \in \mathbf{Z}/M\mathbf{Z}$ and determining the eigenvalues of the spin Hamiltonian completely. The corresponding eigenvalue of the continuum M-particle operator is given by [80]

$$
\mathcal{E}_M(p) = \frac{2M(M-1)}{N} \zeta_N \left(\frac{N}{2} \right) + \sum_{\beta=1}^{M} p_\beta^2/2
$$

$$
- \frac{1}{2} \left[\sum_{k<l}^{m} (2\delta_{c(k),c(l)} F(t_k - t_l) - \delta_{|c(k)-c(l)|,1} F(t_k - t_l)) \right.
$$

$$
\left. - M \sum_{c(k)=1} F(t_k) \right],
$$

where

$$
F(t) = -\wp_N(t) + (\zeta_N(t) - 2/N\zeta_N(N/2))^2 + 4/N\zeta_N(N/2).
$$

It is worth noting that for real calculation of the eigenvalues one has to solve the Bethe-type equations (3.113), (3.116) and (3.117) at first. It is not clear how to treat properly this huge system of highly transcendental equations even in the limit $N \to \infty$. In this limit, there is a procedure known as asymptotic Bethe Ansatz (ABA) which consists in imposing periodic boundary condition on the asymptotics of the wave functions for infinite lattice [50]. It will be used in the next subsection for obtaining some results on antiferromagnetic ground state.

3.4.4. *ABA Results for Large N*

In this subsection, the ABA hypothesis (still unproved) will be used for description of some properties of the spin chain with the exchange

$$h(j) = \frac{\sinh^2 \kappa}{\sinh^2 \kappa(j-k)}, \tag{3.118}$$

which corresponds to $J = -(\sinh \kappa/\kappa)^2$ in (3.1) (the antiferromagnetic regime) at large but finite N. Note that in the nearest-neighbor limit $\kappa \to \infty$, one can decompose (3.1) with the exchange (3.118) as

$$H = \frac{1}{2} \sum_j (\vec{\sigma}_j \vec{\sigma}_{j+1} - 1) + \frac{1}{2} e^{-2\kappa} \sum_j (\vec{\sigma}_j \vec{\sigma}_{j+2} - 1) + o(e^{-2\kappa}).$$

Hence, one can write the ground-state energy per site as

$$e = \frac{1}{2} \langle \vec{\sigma}_j \vec{\sigma}_{j+1} - 1 \rangle + \frac{1}{2} e^{-2\kappa} \langle \vec{\sigma}_j \vec{\sigma}_{j+2} - 1 \rangle + o(e^{-2a}), \tag{3.119}$$

where $\langle\ \rangle$ means average on the vacuum state of the Hamiltonian. Fortunately, in the first order approximation one can replace this state to the vacuum state of non-perturbed Hamiltonian with the interaction of nearest-neighbor spins, $H_0 = \frac{1}{2} \sum_j (\vec{\sigma}_j \vec{\sigma}_{j+1} - 1)$. It gives an opportunity to find the second-neighbor correlator $\vec{\sigma}_j \vec{\sigma}_{j+2}$ by calculating (3.119) explicitly.

The scheme of ABA is based on asymptotic expression of the wave function with M down spins for infinite chain in the region $n_1 \ll n_2 \cdots \ll n_M$, which has been described in Section 3.3,

$$\psi(n_1, \ldots, n_M) \propto \sum_{P \in \pi_M} \exp\left(i \sum_{\alpha=1}^{M} k_{P\alpha} n_\alpha\right) \exp\left(\frac{i}{2} \sum_{\alpha<\beta}^{M} \chi(p_{P\alpha}, p_{P\beta})\right),$$

where the first sum is taken over all permutations from the group π_M, $\{p_\alpha\}$ is the set of pseudomomenta and $\chi(p_\alpha, p_\beta)$ is the two-magnon phase shift defined by the relations

$$\cot \frac{\chi(p_\alpha, P_\beta)}{2} = \varphi(p_\alpha) - \varphi(p_\beta),$$

$$\varphi(p) = \frac{p}{2\pi i \kappa} \zeta_1\left(\frac{i\pi}{2\kappa}\right) - \frac{1}{2i\kappa} \zeta_1\left(\frac{ip}{2\kappa}\right).$$

To consider the chains of finite length N in the thermodynamic limit $N \to \infty$, we adopt the main hypothesis of ABA, i.e. imposing periodic boundary conditions on the asymptotic form of the wave function. Taking $\psi(n_2, \ldots, n_M, n_1 + N) = \psi(n_1, \ldots, n_M)$ and calculating both sides with the use of the above formula for asymptotic results in the ABA equations

$$\exp(ip_\alpha N) = \exp\left(i \sum_{\beta \neq \alpha}^{M} \chi(p_\alpha, p_\beta) \right), \quad \alpha = 1, \ldots, M. \tag{3.120}$$

The energy of corresponding configuration is given by

$$E_M = \sum_{\alpha=1}^{M} \sum_{n \neq 0} \frac{\sinh^2 a}{\sinh^2 an} (\cos(k_\alpha n) - 1).$$

For investigating the antiferromagnetic vacuum of the model, one should take N even, $M = N/2$. Taking logarithms of both sides of (3.120) and choosing the proper branches, one arrives at

$$\frac{Q_\alpha}{N} = \frac{\pi - p_\alpha}{2\pi} - \frac{1}{\pi N} \sum_{\beta \neq \alpha}^{M} \arctan \left[\varphi(p_\alpha) - \varphi(p_\beta) \right],$$

where $\{Q\}$ is the set of (half) integers. For antiferromagnetic ground state, one assumes as usually that these numbers form a uniform string from $-Q_{\max}$ to $Q_{\max}$, $Q_{\max} = N/4 - 1/2$ without holes. After introducing rapidity variable λ by the relation $\lambda = \varphi(k)$ and the function $\mu(\lambda)$ via the relation $\pi - k = \mu(\lambda)$, the ABA equations (3.120) can be written as [64]

$$Q_\alpha/N = Z(\lambda \alpha), \tag{3.121}$$

where

$$Z(\lambda) = (2\pi)^{-1} \mu(\lambda) - \frac{1}{\pi N} \sum_{\beta=1} \arctan(\lambda - \lambda_\beta).$$

Following [64], let us go to the continuous variable $x = Q_\alpha/N$ in the limit $N \to \infty$ and introduce the root density $\sigma_N(\lambda)$ by the relation

$\sigma_N(\lambda) = dx/d\lambda$. Differentiating both sides of (3.120) with respect to λ, one arrives at the following equation in the limit $N \to \infty$

$$\sigma_\infty(\lambda) = (2\pi)^{-1}\mu'(\lambda) - \int_{-\infty}^{\infty} A(\lambda - \lambda')\sigma_\infty(\lambda')d\lambda', \qquad (3.122)$$

where $A(\lambda) = [\pi(1 + \lambda^2)]^{-1}$. The energy per site can be written as

$$e_\infty = \lim_{N\to\infty} N^{-1}E_{N/2} = \int_{-\infty}^{\infty} \varepsilon(p(\lambda))\sigma_\infty(\lambda)d\lambda, \qquad (3.123)$$

where

$$\varepsilon(p(\lambda)) = 2\sinh^2\kappa \sum_{n=1}^{\infty} \frac{\cos np(\lambda) - 1}{\sinh^2\kappa n}.$$

The solution to (3.122) can be found via Fourier transform as follows:

$$\sigma_\infty(\lambda) = (2\pi)^{-2} \int_{-\infty}^{\infty} \frac{e^{i\lambda k}dk}{1 + e^{-|k|}} \int_{-\infty}^{\infty} \mu'(\tau)e^{-ik\tau}d\tau.$$

Substituting it into (3.123) yields

$$e_\infty = (2\pi)^{-2} \int_{-\infty}^{\infty} d\lambda\varepsilon(p(\lambda)) \int_{-\infty}^{\infty} dk\frac{e^{ik\lambda}}{1 + e^{-|k|}} \int_{-\infty}^{\infty} \mu'(\tau)e^{-ik\tau}d\tau.$$

Upon choosing variables as $\lambda = \varphi(p), \mu'(\tau)d\tau = -dp'$ and changing the order of integration (it is allowed since the integral over τ vanishes sufficiently fast as $|k| \to \infty$), one arrives at the following formula for an energy per site:

$$e_\infty = -(2\pi)^{-2} \int_{-\infty}^{\infty} \frac{dk}{1 + e^{-|k|}} \int_0^{2\pi} dp\varepsilon(p)\varphi'(p)e^{ik\varphi(p)}$$

$$\times \int_0^{2\pi} dp'e^{-ik\varphi(p')}, \qquad (3.124)$$

where the functions $\varepsilon(p)$ and $\varphi(p)$ are determined as above. Unfortunately, the integrals in (3.124) cannot be evaluated analytically; however, one can find as $\kappa \to \infty$ that

$$\varphi(p) = \frac{1}{2}\cot\frac{p}{2} + 2e^{-2\kappa}\sin p + o(e^{-2\kappa}),$$

$$\varepsilon(p) = 2(\cos p - 1) + 2e^{-2\kappa}(\cos 2p - 1) + o(e^{-2\kappa}).$$

Upon substituting these expressions into (3.124), the inner integrals are calculated analytically up to the order $e^{-2\kappa}$ and final result for second-neighbor correlator in the model with nearest-neighbor exchange reads

$$\langle \vec{\sigma}_j \vec{\sigma}_{j+2} \rangle = 1 - 16 \ln 2 + 9\zeta(3),$$

where ζ is the Riemann zeta function which appears in the right-hand side due to the formula $\int_0^\infty \frac{k^2 dk}{1+e^k} = 3/2\zeta(3)$. This result coincides exactly with the expression given by Takahashi [78] who considered the limit of infinite one-site repulsion in the half-filled Hubbard model.

Another ABA result is the calculation of *central charge* c of underlying conformal field model [64]. It is given by the formula for finite-N correction to the energy of antiferro-magnetic ground state

$$\Delta e_N = e_N - e_\infty = -\frac{\pi c}{6N^2}\xi,$$

where ξ is the velocity of the lowest-lying elementary excitations. The value of Δe_N can be calculated via Equations (3.121) where the values of the order N^{-2} should be carefully taken into account. We would like to mention only the final result of rather long calculations [64],

$$\Delta_{eN} = -(12N^2)^{-1}\phi_\infty + O(N^{-3}),$$

$$\phi_\infty = 2\pi i \lim_{\lambda \to \infty} \frac{\int_{-\infty}^\infty kdk \frac{\exp(ik\lambda)}{1+\exp(-|k|)} \int_0^{2\pi} dp\,\varepsilon(p)\varphi'(p)\exp[-ik\varphi(p)]}{\int_{-\infty}^\infty dk \frac{\exp(ik\lambda)}{1+\exp(-|k|)} \int_0^{2\pi} dp \exp[-ik\varphi(p)]}.$$

The energy and momentum of elementary excitations over antiferro-magnetic vacuum can also be calculated on the base of (3.121) under an assumption that this excitation corresponds to presence of a hole in the sequence of numbers $\{Q\}$. These calculations result in the formula $\xi = (2\pi)^{-1}\phi_\infty$ which gives the value of the central charge $c = 1$ as in the case of usual nearest-neighbor chain.

3.5. Inhomogeneous Lattices

It is generally believed that more general dynamical Calogero–Moser systems describing particles with internal degrees of freedom are integrable. The motion of particles can be eliminated by arranging them into classical equilibrium positions. By this way, the first model of *inhomogeneous* chain [68] has been obtained where spin interaction was given by inverse squares of distance between them and spins were located on equilibrium positions of particles with rational two-body interaction in the field with a harmonic potential. As for inverse hyperbolic square exchange, the integrability of the corresponding models is still questionable. Anyway, there are many indications to this fact as it will be shown later.

The integrability of classical Calogero–Moser systems in some external fields has been considered in [76]. It was shown there that the Hamiltonians (3.3) with interaction (3.5) (with $\kappa = 1$ as it can be removed by scaling transformation) are still integrable if the external field with the potential

$$W(x) = \alpha^2 \cosh(4x) + 2\beta \cosh(2x) + 2\gamma \sinh(2x) \qquad (3.125)$$

is added. As for spin chains, the Hamiltonian is still given by

$$H = \sum_{j<k}^{N} h_{jk} P_{jk}, \qquad (3.126)$$

where $\{P_{jk}\}$ is any representation of the symmetric group π_N, $h_{jk} = \sinh^{-2}(x_j - x_k)$ and $\{x_j\}$ are the coordinates of classical particles at equilibrium obeying the equations

$$-2 \sum_{k \neq j} h_{jk} c_{jk} + W'(x_j) = 0, \qquad (3.127)$$

where

$$c_{jk} = \coth(x_j - x_k).$$

The first question is to construct the Lax pair for these systems. Consider the following Ansatz of $(2N \times 2N)$ matrices (L, M) with entries

$$L^{11} = -L^{22} = L_0, \quad L^{12} = L^{21} = \psi + \rho, \quad M^{11} = M^{22} = M_0 + m,$$
$$M^{12} = M^{21} = \phi,$$

where L_0 and M_0 form the standard Lax pair for the systems without external field,

$$(L_0)_{jk} = (1 - \delta_{jk})c_{jk}P_{jk}, \quad (M_0)_{jk} = (1 - \delta_{jk})h_{jk}P_{jk} - \delta_{jk}\sum_{s \neq j}^{N} h_{js}P_{js}$$

and ψ, ϕ, ρ and m are $(N \times N)$ matrices with entries

$$(\psi)_{jk} = \xi(z_j)\delta_{jk}, \quad \phi_{jk} = \varphi(z_j)\delta_{jk}, \quad (m)_{jk} = \mu(z_j)\delta_{jk},$$
$$(\rho)_{jk} = (1 - \delta_{jk})P_{jk},$$

where $z_j = \exp(2x_j)$. The Lax relation $[H, L] = [L, M]$ is equivalent to the set of functional equations

$$c_{jk}[\mu(z_j) - \mu(z_k)] + [\varphi(z_j) + \varphi(z_k)] = 0,$$
$$c_{jk}[\varphi(z_j) + \varphi(z_k)] + h_{jk}[\xi(z_j) - \xi(z_k)] + \mu(z_j) - \mu(z_k) = 0.$$

The general solution to this set is given in [77],

$$\mu(z) = \mu_1 z + \mu_2 z^{-1}, \quad \varphi(z) = -\mu_1 z + \mu_2 z^{-1},$$
$$\xi(z) = \mu_1 z + \mu_2 z^{-1} + \gamma.$$

The potential of an external field reads

$$W(z) = 2[\mu_1^2 z^2 + \mu_2^2 z^{-2} + (2_\gamma - 1/2)(\mu_1(z) + \mu_2 z^{-1})].$$

It contains three free parameters as (3.125). For the special case of the external Morse potential ($\mu_2 = 0$), the matrix M obeys also the condition $\sum_{j=1}^{2N} M_{jk} = 0$, which guarantees that the integrals of motion can be constructed as $\{\sum_{j,k}^{2N}(L^n)_{jk}\}$. In other cases, the existence of the Lax pair does not imply integrability immediately.

The extra integrals of motion should be some polynomials in the permutations as it takes place for usual lattice spin models [73]. It turns out that minimal degree of this polynomial now equals 3 and the operator

$$I = \sum_{j \neq k \neq l \neq m}^{N} c_{jk} c_{kl} P_{jk} P_{kl} P_{lm} - \frac{1}{2} \sum_{j \neq k \neq l} (c_{jl} - c_{kl})^2 P_{jk}$$

$$+ \sum_{j \neq k}^{N} [F(x_j) + F(x_k)] P_{jk}$$

commutes with H if F is a solution of functional equation

$$g(x_j, x_k) + g(x_k, x_l) + g(x_l, x_j) = 0,$$

where

$$g(x_j, x_k) = 2h_{jk}(F(x_j) - F(x_k)) + c_{jk}(W'(x_j) + W'(x_k)).$$

The solution is given by the relation $g(x_j, x_k) = G(x_j) - G(x_k)$ and functional equation for the potential

$$c_{jk}(W'(x_j) + W'(x_k)) - 2h_{jk}(W(x_j) - W(x_k)) = G(x_j) - G(x_k).$$

Its general solution just gives the form (3.125), which supports the hypothesis of complete integrability of this class of models.

To construct the explicit eigenvalues of the corresponding spin Hamiltonians, one needs more knowledge about the solutions to equilibrium equations (3.127). It can be easily done for special case of the Morse potential $W(x) = 2\tau^2(\exp(4x) - 2\exp(2x))$, where these equations have the form [77]

$$-\sum_{k \neq j}^{N} \frac{z_k(z_j + z_k)}{(z_j - z_k)^3} + \tau^2(z_j - 1) = 0, \qquad (3.128)$$

where the variable $z = \exp(2x)$ is introduced. Following the observation of Calogero [69], one can assume that the roots $\{z_j\}$ of (3.128) are given by roots of some polynomial $p_N(z) = \prod_{j=1}^{N}(z - z_j)$ obeying the second-order differential equation. In the case of the Morse potential, this equation reads

$$y \frac{d^2 p_N(y)}{dy^2} + (-y + \Gamma + 1) \frac{d p_N(y)}{dy} + N p_N(y) = 0, \qquad y = 2\tau z,$$

where $\Gamma = 2(\tau - N) + 1$. It means that p_N are the well-known Laguerre polynomials $L_N^{(\Gamma)}(2\tau z)$. The following properties of their roots will be used:

(i) For $\Gamma > -1$, all roots of L are real positive numbers.
(ii) As $\Gamma = -N + \varepsilon, \varepsilon \to 0$, all the roots of L approach 0 with the asymptotic behavior

$$z_j \sim \mathrm{const}|\varepsilon|^{1/N} \exp\left(\frac{2\pi i j}{N}\right).$$

The rational Calogero spin chain with inverse square exchange [68] is obtained as a limit of $\tau \to \infty, z_j = 1 + \tau^{-1/2}\xi_j$. The lattice points in this limit are the roots of the Hermite potential $H_N(\xi)$. As $\Gamma \to N$, the lattice becomes equidistant in angles and the model upon rescaling is just the trigonometric Haldane–Shastry model [67]. Hence, the inhomogeneous model defined by the lattice (3.128) can be considered as interpolating between the Haldane–Shastry and Polychronakos models.

If one chooses as $\{P_{jk}\}$ in (5.2) the spin representation of the permutation group, $P_{jk} = (1 + \vec{\sigma}_j\vec{\sigma}_k)/2$, the eigenvectors can be treated as in Sections 3.4 and 3.5. That is, one can start from the ferromagnetic vacuum $|0\rangle$ with all spins up and consider the states with given number of down spins M,

$$|\psi^{(M)}\rangle = \sum_{n_1 \neq n_2 \cdots \neq n_M}^{N} \psi(n_1, \ldots, n_M) \prod_{s=1}^{M} \sigma_s^- |0\rangle.$$

With the use of the properties of the Laguerre polynomials, one finds that in one-magnon sector the wave functions can be represented as

$$\psi_m(n) \propto z_n^m \frac{L_{N-m-1}^{(\Gamma+2m)}(2\tau z_n)}{L_{N-1}^{(\Gamma)}(2\tau z_n)}, \quad m = 0, \ldots, N-1.$$

The corresponding energies up to universal constant $C_N = N(N-1)(3\Gamma + 2N - 1)/24$ are given by

$$E_m^{(1)} = \epsilon_m = -\frac{m}{2}(\Gamma + m).$$

The two-magnon wave functions can be found analytically and the complete set of $N(N-1)/2$ eigenvalues can be written as

$$E^{(2)}_{m,n} = \epsilon_m + \epsilon_n(1 - \delta_{m,n-1}), \quad 0 \le m < n \le N - 1.$$

In the M-magnon sector, one can find analytically only some eigenstates within the Ansatz

$$\psi(n_1, \ldots, n_M) = \frac{\prod_{\lambda < \mu}^M (z_{n\lambda} - z_{n\mu})^2}{\prod_{v=1}^M p'_N(z_{nv})} F(z_{n1}, \ldots, z_{nM}),$$

where F is some symmetric polynomial in $\{z\}$. It comprises $(N - M + 1)![M!(N - 2M + 1)!]^{-1}$ eigenvalues which are still additive,

$$E^{(M)}_{\{m_k\}} = \sum_{k=1}^M \epsilon_{m_k}, \quad m_k < m_{k+1} - 1, \quad 0 \le m_k \le N - 1.$$

This formula allows one to make the hypothesis about the structure of the whole set of eigenvalues which are described by

$$E_{l_1, \ldots, l_k} = \sum_{k=1}^{N-1} \epsilon_k l_{k+1}(1 - l_k),$$

where $\epsilon_k = -k(\Gamma + k)/2$ and $\{l_k\} = 0, 1$. As a consequence of this hypothesis, the Hamiltonian $H = 2 \sum_{j<k}^N h_{jk} \vec{\sigma}_j \vec{\sigma}_k$ is unitary equivalent to the Hamiltonian of the classical one-dimensional Ising model with non-uniform magnetic field,

$$H_I = \epsilon_{N-1} \sigma_N + \sum_{k=0}^{N-2} [\sigma_{k+1}(\epsilon_k - \epsilon_{k+1}) - \sigma_{k+1}\sigma_{k+2}\epsilon_{k+1} \quad (3.129)$$

with $\{\sigma_k\} = \pm 1$. This result comprises the two above analytical formulae for the spectrum as well as Haldane–Shastry and harmonic limits and is confirmed by numerical diagonalization of small lattices up to $N = 12$ with several values of the parameter τ.

The simplicity of the spectrum (3.129) allows one to compute the free energy f as a function of the inverse temperature β in the thermodynamic limit upon rescaling the magnon energies with a factor N^{-2} [86]. With the use of quasiparticle dispersion law $\epsilon(x) = -x(\gamma + x)/2$ where $\gamma = \Gamma/N$, one obtains

$$f = -\frac{1}{\beta}\left(\int_0^{-\gamma} dx \log[1 + \exp(\beta\epsilon(x))]\right.$$
$$\left. + \int_{-\gamma}^1 dx \log[1 + \exp(-\beta\epsilon(x))]\right),$$

which gives at $\gamma = -1$ the result exactly coinciding with the free energy of the trigonometric Haldane–Shastry model.

Coming back to the general potential of an external field (3.125), one has to start with the equilibrium equations

$$-\sum_{k\neq j}^N \frac{z_k(z_j + z_k)}{z_j - z_k^3} + \frac{1}{4}\sum_{j=1}^N[\alpha^2(z_j - z_j^{-3}) + \beta + \gamma - (\beta - \gamma)z_j^{-2}] = 0.$$

$$(3.130)$$

As in the case of the Morse potential described above, let us introduce the polynomial

$$p_N(z) = \prod_{j=1}^N (z - z_j)$$

with the use of the solutions to (3.130) and try to identify the differential equation which this polynomial might satisfy. To do this, note that the function $F_j(z) = z(z + z_j)(z - z_j)^{-3}d\log p_N(z)/dz$ has simple poles at $z = z_k$ with proper residues, and the equilibrium equations can be recast in the form

$$\mathrm{res}F_j(z)|_{z=z_j} = 2a_{1j} + z_j(4a_{2j} - 3a_{1j}^2) + z_j^2(a_{3j} + a_{1j}^3 - 2a_{1j}a_{2j})$$
$$= \alpha^2(z_j - z_j^{-3}) + \beta + \gamma - (\beta - \gamma)z_j^{-2}, \qquad (3.131)$$

where $a_{\lambda j} = [p_N'(z_j)]^{-1}(d/dz)^{\lambda+1}p_N(z)|_{z=z_j}$. If one supposes that $p_N(z)$ obeys the second-order differential equation

$$z^2 p_N''(z) + w_1(z)p_N'(z) + w_2(z)p_N(z) = 0 \qquad (3.132)$$

with some polynomials $w_{1,2}(z)$, one finds upon consecutive differen-tiations of (3.132) with the use of the formula $p_N(z_j) = 0$ that the equilibrium equations in the form (3.131) are equivalent to

$$\frac{d}{dz}\left[w_2 + \frac{1}{4}(\alpha^2(z^2 + z^{-2}) - z^{-2}w_1^2) + \frac{1}{2}w_1'\right.$$

$$\left. + 2(z(\beta + \gamma) + (\beta - \gamma)z^{-1})\right] = 0.$$

This condition is satisfied by $w_1(z) = -\alpha(z^2 - 1) + (4\alpha^{-1}\beta - \gamma_1)z$, $w_2(z) = (\alpha - 4\beta)z + e_N$, where $\gamma_1 = 4\alpha^{-1}\gamma$ and parameter e_N is still unknown. One of the solutions to (3.132) is a polynomial of the degree N if the parameters α and β are restricted by

$$\beta = -\frac{N-1}{4}\alpha.$$

Equation (3.132) is now written as

$$z^2 p_N''(z) - [\alpha(z^2 - 1) + (\gamma_1 + N - 1)z]p_N'(z) + (\alpha N z + e_N) = 0. \tag{3.133}$$

The substitution $p_N(z) = z^N + \sum_{l=0}^{N-1} d_l z^l$ results in the recurrence relation for d-coefficients in the form

$$\alpha d_{l-1}(N - l + 1) + d_l[e_N + l(l - \gamma_1 - N)]$$

$$+ \alpha(l + 1)d_{l+1} = 0, \quad l = 0, \ldots, N.$$

It should be solved under the boundary conditions

$$d_1 = 0, \quad d_N = 1, \quad d_{N+1} = 0.$$

The last condition results in Nth order equation for the parameter e_N. The solution must be chosen so as to have all the roots of $p_N(z)$ positive. It is unique since the system of particles which repel each other has only one equilibrium point being confined in the field with potential (3.125).

Due to (3.133), various symmetric combinations of the roots of (3.130) can be expressed analytically in terms of α, γ_1 and e_N. In particular, the energy of classical equilibrium configuration does not depend on e_N and is given by

$$E_{cl} = -\frac{N}{2}\left(\frac{N^2 - 1}{3} + \gamma_1^2 - 2\alpha^2\right).$$

As for corresponding spin chain with the Hamiltonian $H = \sum_{j<k} h_{jk}(\vec{\sigma}_j \vec{\sigma}_k - 1)$, the strategy for finding eigenvalues is the same as for the Morse potential described above. However, the information which could be obtained this way is much more scarce. In M-magnon sectors with $M \leq N/2$, one can use the Ansatz

$$\psi(n_1, \ldots, n_M) = \frac{\prod_{\lambda>\mu}^{M}(z_{n_\lambda} - z_{n_\mu})^2}{\prod_{\mu=1}^{M} p'_N(z_{n_\mu})} Q(z_{n_1}, \ldots, z_{n_M})$$

for multimagnon wave function, and show that the eigenequation can be cast in the form

$$\sum_{j=1}^{M} \left\{ z_j^2 \frac{\partial^2}{\partial z_j^2} - [\alpha(z_j^2 - 1) + (\gamma_1 + N - 3)z_j] \frac{\partial}{\partial z_j} \right.$$

$$+ 2 \sum_{j \neq k}^{M} \frac{z_j^2 \partial/\partial z_j - z_k^2 \partial/\partial z_k}{z_j - z_k} + M[(M-1)(4M+1)/3$$

$$\left. - M(\gamma_1 + N - 1) + e_N] + \alpha(N - 2M) \sum_{k=1}^{M} z_k - 2E_M \right\} Q = 0.$$

For even N, the solution at $M = N/2\,(S_z = 0)$ is given by $Q =$ constant and the corresponding eigenenergy reads

$$E_{N/2} = 1/2\{M[(M-1)(4M+1)/3 - M(\gamma_1 + N - 1) + e_N]\}.$$

It was verified numerically that for small lattices ($N \leq 8$) at various sets of parameters α and γ_1 this is the exact ground state of the antiferromagnetic chain (3.126). Unfortunately, this approach does not allow to identify other states and write down such a simple formula for the whole spectrum as in the case of the Morse potential.

3.6. The Related Hubbard Chains: Are They Integrable?

There are other many-body systems on a lattice connected to the Heisenberg–Dirac–van Vleck spin chains discussed above: the itinerant fermions of spin 1/2 which interact being at the same lattice site. The corresponding models are Hubbard chains with the Hamiltonian

$$H_{Hub} = \sum_{j \neq k; \sigma}^{N} t_{jk} c_{j\sigma}^{+} c_{k\sigma} + 2U \sum_{j}^{N} (c_{j\uparrow}^{+} c_{j\uparrow} - 1/2)(c_{j\downarrow}^{+} c_{j\downarrow} - 1/2), \quad (3.134)$$

where the operators $c_{j\sigma}^{+}$ create fermion with spin σ on the site j,

$$\{c_{j\sigma}^{+}, c_{k\sigma'}\} = \delta_{jk} \delta_{\sigma\sigma'}, \quad \{c_{j\sigma}, c_{k\sigma'}\} = 0, \quad (3.135)$$

$t_{jk} \equiv t(j - k)$ is the hopping matrix comprising probability amplitudes for hopping between sites j and k (it is supposed to be Hermitian) and $U > 0$ is the strength of on-site repulsion.

This model was originally introduced by J. Hubbard [83] in three dimensions to describe a metal-insulator transition for systems of fermions with spin. It was found that the one-dimensional version (3.134) is solvable by the Bethe Ansatz [84] in the case of nearest-neighbor hopping under periodic boundary conditions,

$$t(j) = \delta_{|j|,1} + \delta_{|j|,N-1}. \quad (3.136)$$

The proof of integrability of (3.134) with the hopping (3.136), i.e. constructing of the non-trivial integrals of motion which commute with (3.136), came much later [85]. There are two trivial invariants: total number of fermions M and number of fermions of up (down) spins which are conserved due to $su(2)$ invariance of (3.134).

The connection with Heisenberg–Dirac–van Vleck chains discussed above comes in the limit of infinite U at $M = N$ (half-filled band). In this limit, fermions are not allowed to occupy the site twice and hop, i.e. they can interact only via spin exchange. The spin Hamiltonian, which arises in the lowest order in t/U, has the form

$$H_{spin} = \sum_{j \neq k}^{N} |t_{jk}|^2 \vec{\sigma}_j \vec{\sigma}_k. \quad (3.137)$$

It is this relation on which Gebhard and Ruckenstein (GR) [86] proposed the solvable model with hopping

$$t(j) = \frac{N}{\pi} \frac{1}{\sin\left(\frac{\pi j}{N}\right)}. \tag{3.138}$$

They were able to guess the simple effective Hamiltonian which comprises all the spectrum of H_{Hub} with hopping (3.138) but failed in proving this result analytically. Note that till now this proof is lacking despite the physical consequences of the GR hypothesis being investigated thoroughly [87] and numerical calculations also supporting it. Moreover, on the base of (3.137), yet another model has been proposed [88] with short-range hopping on the infinite lattice,

$$t(j) = -i \sinh \kappa / \sinh(\kappa j). \tag{3.139}$$

The authors of [88] used the hypothesis of the asymptotic Bethe Ansatz for the model (3.139) without any proof of integrability and found quite satisfactory properties in the thermodynamic limit. They showed also that (3.139) includes, as a limit of $\kappa \to \infty$, the nearest-neighbor hopping (3.136) on the infinite lattice.

On the base of correspondence with H_{Hub} and its limit (3.137), one can guess also the integrability of the elliptic model with hopping being some "square root" of elliptic exchange (3.12). But in all these cases, one has to find conserved quantities so as to prove integrability without appeal to any limit or numerical calculations. This problem has not been solved completely to date. But some explicit indications to the integrability are found and will be discussed later.

In the spectrum of the model with long-range hopping (3.138), some degeneracies were found similar to the degeneracies for the Haldane–Shastry model [89]. This shows that the model might have additional symmetry besides the usual one. For the Haldane–Shastry model, it was found that this symmetry is given by infinite vector algebra, the $sl(2)$ Yangian discovered before in [90]. It is natural to try to find at first the source of degeneracies for the Gebhard–Ruckenstein model (3.138). Due to explicit $sl(2)$-invariance of the Hubbard Hamiltonian, it is useful to introduce, instead of fermion

c-operators, their bilinear spin-like combinations extending the concept of spin to different sites. That is, the product of operators $c_{j\sigma}^{+} c_{k\tau}$ can be arranged as a 2×2 matrix $(S_{jk})_{\tau}^{\sigma}$ labeled by spin indices, which allows one to define the S-operators as

$$S_{jk}^{\alpha} = \operatorname{tr}(\sigma^{*\alpha} S_{jk}), \quad S_{jk}^{0} = \operatorname{tr}(S_{jk}), \quad S_{j}^{\alpha} = S_{jj}^{\alpha}, \quad S_{j}^{0} = S_{jj}^{0},$$

where σ_{α} are the Pauli matrices. Note that $S_{j}^{\alpha}/2$ and S_{j}^{0} are the spin density and fermion density operators. The commutators of these S-operators are

$$[S_{jk}^{0}, S_{lm}^{0}] = \delta_{kl} S_{jm}^{0} - \delta_{mj} S_{lk}^{0},$$

$$[S_{jk}^{0}, S_{lm}^{\alpha}] = \delta_{kl} S_{jm}^{\alpha} - \delta_{mj} S_{lk}^{\alpha}, \quad (3.140)$$

$$[S_{jk}^{\alpha}, S_{jk}^{\beta}] \delta^{\alpha\beta} (\delta_{kl} S_{jm}^{0} - \delta_{mj} S_{lk}^{0}) + i\varepsilon^{\alpha\beta\gamma} (\delta_{kl} S_{jm}^{\gamma} + \delta_{mj} S_{lk}^{\gamma}) .$$

There are a lot of other relations between these operators due to their composite nature. Some of them can be written down explicitly as follows:

$$S_{jk}^{\alpha} S_{lm}^{\alpha} + S_{jk}^{0} S_{lm}^{0} + 2 S_{jm}^{0} S_{lk}^{0} = 4 \delta_{kl} S_{jm}^{0} + 2 \delta_{lm} S_{jk}^{0},$$

$$S_{jk}^{\alpha} S_{lm}^{\alpha} + S_{lm}^{0} S_{jk}^{\alpha} + S_{lk}^{0} S_{jm}^{\alpha} + S_{jm}^{0} S_{lk}^{\alpha} = \delta_{jk} S_{lm}^{\alpha}$$

$$+ \delta_{lm} S_{jk}^{\alpha} + \delta_{lk} S_{jm}^{\alpha} + \delta_{jm} S_{lk}^{\alpha},$$

$$S_{jk}^{\alpha} S_{lm}^{\beta} + S_{jk}^{\beta} S_{lm}^{\alpha} + S_{jm}^{\alpha} S_{lk}^{\beta} + S_{jm}^{\beta} S_{lk}^{\alpha} = \delta^{\alpha\beta} (S_{jm}^{0} (2\delta_{lk} - S_{lk}^{0})$$

$$+ S_{jm}^{\gamma} S_{lk}^{\gamma}), -i\varepsilon^{\alpha\beta\gamma} S_{jk}^{\beta} S_{lm}^{\gamma} - S_{jm}^{0} S_{lk}^{\alpha} + S_{lk}^{0} S_{jm}^{\alpha} = 2\delta_{lk} S_{jm}^{\alpha}$$

$$+ \delta_{jk} S_{lm}^{\alpha} - \delta_{lm} S_{jk}^{\alpha}. \quad (3.141)$$

These basic relations contain also a whole list of others which appear upon equating all possible combinations of site indices. In terms of S-operators, the Hubbard Hamiltonian reads

$$H_{\text{Hub}} = \sum_{j \neq k} t_{jk} S_{jk}^{0} + U \sum_{j} ((S_{j}^{0} - 1)^2 - 1/2). \quad (3.142)$$

The operators of total spin $I^{\alpha} = 1/2 \sum_{j} S_{j}^{\alpha}$ commute with (3.142), their sl_2 commutation relations are obtained from (3.140) by summation over lattice sites, $[I^{\alpha}, I^{\beta}] = i\varepsilon^{\alpha\beta\gamma} I^{\gamma}$. Consider now the operator

$$J^\alpha = \frac{1}{2} \sum_{j \neq k} ((f_{jk} + h_{jk}(S_j^0 + S_k^0 - 2))S_{jk}^\alpha + g_{jk}\varepsilon^{\alpha\beta\gamma}S_j^\beta S_k^\gamma), \quad (3.143)$$

where $f_{jk} \equiv f(j-k)$, etc. and g and h are odd functions. It is possible to show, with the use of (3.140) and (3.141), that H_{Hub} commutes with J_α if the following set of functional equations is satisfied [91]:

$$(g_{jl} - g_{kl})h_{jk} = \tfrac{i}{2}h_{jl}h_{kl}, \quad j \neq k \neq l \neq j,$$

$$Uf_{jk}/2h_0 + g_{jk}h_{jk} = -\tfrac{i}{4}\sum_l h_{jl}h_{kl}, \quad j \neq k ,$$

$$\sum_l (f_{jl}h_{kl} - f_{kl}h_{jl}) = 0,$$

$$t_{jk} = h_0 h_{jk};$$

where h_0 is a free parameter. It turns out that the only solutions to these equations just give the trigonometric (finite N) and hyperbolic (infinite lattice) forms of hopping (3.138) and (3.140)! In the trigonometric case, one finds

$$f_{jk} = 0, \quad g_{jk} = \frac{1}{2}\cot(\pi(j-k)/N), \quad h_{jk} = i\sin^{-1}(\pi(j-k)/N),$$

whereas in the hyperbolic case,

$$f_{jk} = \frac{\sinh(\kappa)(j-k)}{U\sinh(\kappa(j-k))}, \quad g_{jk} - \frac{1}{2}\coth(\kappa(j-k)),$$

$$h_{jk} = i\sinh^{-1}(\kappa(j-k)).$$

Note that J^α does not depend on U in the trigonometric case. It is natural to ask to which symmetry does this new vector operator correspond. It turns out that this symmetry is just Yangian $Y(sl_2)$ as it can be seen from the commutation relations

$$[I^\lambda, J^\mu] = i\varepsilon_{\lambda\mu\nu}J^\nu,$$

$$[J^\alpha, K^\beta] + [J^\beta, K^\alpha] = 0, \quad (3.144)$$

where

$$K^\alpha = i\varepsilon^{\alpha\beta\gamma}[J^\beta, J^\gamma] - 4\delta(I^\beta)^2 I^\alpha$$

and $\delta = -1$ in the trigonometric case and 1 in hyperbolic one. Equation (3.144) is just the defining relation for sl_2 Yangian.

Note also that for all odd functions $t(j)$ there is a canonical transformation

$$c_{j\downarrow} \to c_{j\downarrow}, \quad c_{j\uparrow} \to c_{j\uparrow}^+, \quad U \to -U, \tag{3.145}$$

which leaves the Hamiltonian invariant but transforms the Yangian generators I^α, J^β into an independent set of generators I'^α, J'^β of another representation of sl_2 Yangian. It turns out that these two representations commute and can be combined to a $Y(sl_2) \oplus Y(sl_2)$ double Yangian. The fact of this commutativity is non-trivial and is of dynamical origin. To verify it and (3.135), one needs the explicit form of the functions f, g, h in (3.143).

The Yangian operator of the nearest-neighbor chain on an infinite lattice found in [92] can be obtained as a limit of the operator (3.142) as $\kappa \to \infty$. In the limit of $U \to \infty$ for the half-filled band, where number of fermions coincides with the number of lattice sites, one can set $S_j^0 = 1$ and recover in the trigonometric case the Yangian for the Haldane–Shastry model [89]. Thus, such rather unlike models as Haldane–Shastry chain and the infinite Hubbard chain with nearest-neighbor hopping are in fact connected: they could be considered as limiting cases of a more general model with the hopping given by elliptic functions.

It is worth noting that the presence of the Yangian symmetry does not imply integrability. To prove integrability, one has to construct the set of *scalar* currents with number of its elements at least equal to the number of lattice sites. It was proved for the Hubbard model with nearest-neighbor hopping by finding its connection to spin ladder and with two coupled six-vertex models [85]. These methods definitely do not work for the Gebhard–Ruckenstein model and its hyperbolic counterpart. One has to find another method for constructing integrals of motion.

To provide examples of the conserved currents which might exist for some choice of the hopping matrix, consider the Ansatz [93]

$$J = \sum_{\substack{j \neq k}}^{N} [A_{jk} S_{jk}^0 + B_{jk}(S_j^0 S_k^0 - \vec{S}_j \vec{S}_k) + D_{jk}(S_j + S_k^0) S_{jk}^0$$

$$+ E_{jk}(S_{jk}^0)^2]. \tag{3.146}$$

which is a most general scalar operator bilinear in $\{S\}$. By definition, $A_{jk} \equiv A(j - k)$, etc. The condition $[H_{\text{Hub}}, J] = 0$ with the use of (3.140) and (3.141) can be cast into the form of two functional equations

$$4t_{jk}(B_{lk} - B_{jl}) + (t_{jl}D_{lk} - D_{jl}t_{lk}) = 0, \qquad (3.147)$$

$$2(t_{jk}E_{kl} + t_{kj}E_{jl}) + (t_{jl}D_{kl} + t_{kl}D_{jl}) = 0, \qquad (3.148)$$

the definition of A being

$$A_{jk} = -2D_{jk} + (2U)^{-1}[-8t_{jk}B_{jk} + 2t_{kj}E_{jk} - r_{jk}],$$

where

$$r_{jk} = \sum_{l \neq j,k}^{N} t_{jl}D_{lk},$$

and several "boundary" equations for t, B and D are as follows:

$$\sum_{l \neq j,k}^{N}(t_{jl}A_{lk} - A_{jl}t_{lk}) = 0,$$

$$\sum_{k \neq j}^{N}(t_{jk}D_{kj} - D_{jk}t_{kj}) = 0,$$

$$\sum_{k \neq j}^{N}(t_{jk}A_{kj} - t_{kj}A_{jk}) = 0.$$

The first functional equation (3.147) is just the Calogero–Moser functional equation with the known general analytic solution. The second functional equation (3.148) always has solutions for E_{jk} if t and D are given by solutions of (3.147). Each function in these and "boundary" equations can be expressed via basic solution to (3.147), and the role of "boundary" equations is to specify the real period of the corresponding Weierstrass functions, which turns out to be N.

The basic solution reads

$$\psi(x) = \frac{\sigma_N(x + \lambda)}{\sigma_N(x)\sigma_N(\lambda)} e^{vx}. \tag{3.149}$$

The other functions in (3.146) and (3.147) are expressed as (recall that $t_{jk} \equiv t(j - k)$ etc.)

$$t(x) = t_0\psi(x), \quad B(x) = -\frac{d}{4}\psi(x)\psi(-x),$$

$$D(x) = d[\psi'(x) - \left(\frac{h\wp'_N(\lambda)}{2} + \zeta_N(\lambda) + v\right)\psi(x)],$$

$$E(x) = \frac{d\psi^2(x)}{2}[1 - h\psi(x + \lambda)\psi(-x - \lambda)],$$

$$r(x) = t_0 d\psi(x)[-(N - 3)\wp_N(x) + h_1(N - 2)\tau(x)$$

$$+(\tau(x) - h_1)(2x\zeta_N(N/2) - N\zeta(x)) + s],$$

$$\tau(x) = \zeta_N(x + \lambda) - \zeta_N(x) - \zeta_N(\lambda), \quad h_1 = h\wp'_N(\lambda)/2,$$

$$s = -(N - 2)\wp_N(\lambda) - \sum_{l=1}^{N-1} \wp_N(l),$$

where σ_N, ζ_N and $\wp_N$ are the Weierstrass elliptic functions determined by the periods $\omega_1 = N, \omega_2 = i\pi/\kappa, \lambda = i\alpha$ or $i\alpha + N/2, v = i\beta, \kappa, d, h, \alpha, \beta$ being arbitrary real parameters. At these conditions, the hopping matrix is Hermitiean. Besides this general solution, there are the degenerate rational, hyperbolic and trigonometric ones, which correspond to one or two infinite periods of the Weierstrass functions. In the first two cases, the lattice should be infinite. Checking the absence of "boundary" terms is a non-trivial task with the key formula

$$[\wp(y + \lambda) - \wp(\lambda)][\zeta(x - y) - \zeta(x + \lambda) + \zeta(y) + \zeta(\lambda)]$$

$$+ [\wp(x + \lambda) - \wp(\lambda)][\zeta(y - x) - \zeta(y + \lambda) + \zeta(x) + \zeta(\lambda)] = \wp'(\lambda).$$

These formulas for t, B, D, E, r, A define the scalar current (3.146) for the model with elliptic hopping which comprises all the hopping matrices (3.136), (3.138), and (3.139) considered above. At λ being the half-period of the Weierstrass $\wp_N$ function, the function $\psi(x)$

becomes odd and yet another current is obtained from (3.146) by the canonical transformation (3.145). It would be of interest to verify by direct calculation that the scalar current commutes with two copies of the Yangian operators in the case of trigonometric and hyperbolic hopping. One can see that trigonometric and hyperbolic models of Bares, Gebhard and Ruckenstein fall into the class of the model with elliptic hopping if one of the periods of the Weierstrass function ω_1 or ω_2, tends to ∞ under an appropriate choice of λ.

It is possible to calculate exact two-electron wave functions for Hamiltonian (3.134) with the general elliptic hopping matrix. The two-electron wave functions for trigonometric and hyperbolic cases were given in the book [94]. The problem is non-trivial only for the $S = 0$ sector of the model. The wave functions for $S = 0$ can be written as

$$\Phi = \sum_{j \neq k} \phi_{jk} c^+_{j\uparrow} c^+_{k\downarrow} |0\rangle + \sum_j g_j c^+_{j\uparrow} c^+_{j\downarrow} |0\rangle,$$

where $|0\rangle$ is the vacuum state. The eigenequation $H\Phi = E\Phi$ can be written in the form

$$\sum_{s \neq j,k} (t_{js}\phi_{sk} + t_{ks}\phi_{js}) + (t_{jk}g_k + t_{kj}g_j) = E\phi_{jk}, \quad (3.150)$$

$$\sum_{k \neq j} t_{jk}(\phi_{jk} + \phi_{kj}) = (E - 2U)g_j. \quad (3.151)$$

In the $S = 0$ sector, ϕ_{jk} is symmetric: $\phi_{jk} = \phi_{kj}$. The proper Ansatz for ϕ_{jk} and g_j reads

$$\phi_{jk} = e^{i(p_1 j + p_2 k)} \varphi(j - k) + e^{i(p_1 k + p_2 j)} \varphi(k - j),$$

$$g_j = g_0 e^{i(p_1 + p_2)j}, \quad (3.152)$$

where

$$\varphi(j) = \varphi_0 \frac{\sigma(j + \tau)}{\sigma(j)}. \quad (3.153)$$

The parameters p_1 and p_2 are connected with τ by the conditions of the periodicity of $\phi_{jk}, \phi_{j+N,k} = \phi_{j,N+k} = \phi_{jk}$,

$$e^{ip_1 N + \eta_1 \tau} = 1, \quad e^{ip_2 N - \eta_1 \tau} = 1, \tag{3.154}$$

where $\eta_1 = 2\zeta(N/2)$. The problem is to find all the parameters $p_{1,2}, g_0$ and τ from eigenequations (3.150) and (3.151) if Ansatz (3.152) and (3.153) are correct. Recall that the elliptic hopping matrix in general is given by

$$t(j-k) = t_0 \psi(j-k) = t_0 e^{\nu(j-k)} \frac{\sigma(j-k+\lambda)}{\sigma(j-k)\sigma(\lambda)}, \tag{3.155}$$

where the factor $\nu = -2\zeta(N/2)\lambda N^{-1}$ is chosen so as to satisfy the periodicity condition, $t(j-k+N) = t(j-k)$. With the use of (3.153) and (3.155), the second eigenequation (3.151) can be cast into the form

$$(E - 2U)g_0 = 2t_0 \varphi_0 S_1(p_1, p_2, \tau), \tag{3.156}$$

where

$$S_1(p_1, p_2, \tau) = \sum_{s \neq 0} e^{\nu s} \frac{\sigma(s+\lambda)}{\sigma^2(s)\sigma(\lambda)} [e^{-ip_2 s} \sigma(s+\tau) + e^{-ip_1 s} \sigma(s-\tau)].$$

The first eigenequation can be written as

$$F(\tau, p_1, p_2, j, k) + F(\tau, p_1, p_2, k, j) + F(-\tau, p_2, p_1, j, k)$$
$$+ F(-\tau, p_2, p_1, k, j) = -g_0[t(j-k)e^{i(p_1+p_2)k}$$
$$+ t(k-j)e^{i(p_1+p_2)j}] + E\phi_{jk}, \tag{3.157}$$

where

$$F(\tau, p_1, p_2, j, k) = \sum_{s \neq j, k} t(j-s)e^{i(p_1 s + p_2 k)}\varphi(s-k)$$
$$= e^{i(p_1 j + p_2 k)} \sum_{q \neq 0, k-j} t(-q)e^{ip_1 q}\varphi(q+j-k).$$

Let us now calculate the sum in the last expression with the use of the explicit forms of $t(j), \varphi(j)$ (Equations (3.145) and (3.147)), and introduce the notation

$$S(l) = \sum_{q \neq 0, l}^{N-1} \frac{\sigma(q - \lambda)\sigma(q - l + \tau)}{\sigma(q)\sigma(\lambda)\sigma(q - l)} e^{(-\nu + ip_1)q}, \qquad (3.158)$$

where $l = k - j \in \mathbf{Z}$, and the function of a continuous argument x,

$$G(l, x) = \sum_{q=0}^{N-1} \frac{\sigma(q - \lambda + x)\sigma(q - l + \tau + x)}{\sigma(q + x)\sigma(\lambda)\sigma(q - l + x)} e^{(-\nu + ip_1)q}. \qquad (3.159)$$

The function $G(l, x)$ is double-quasiperiodic,

$$G(l, x + 1) = e^{\nu - ip_1} G(l, x),$$

$$G(l, x + \omega_2) = e^{\eta_2(\tau - \lambda)} G(l, x),$$

where $\eta_2 = 2\zeta(\omega_2/2)$ (recall that ω_2 is the second period of all the Weierstrass functions here). It has a simple pole at $x = 0$ with decomposition near it of the form

$$G(l, x)|_{x \to 0} = -\frac{\sigma(-l + \tau)}{\sigma(-l)} (x^{-1} + \zeta(-l + \tau) - \zeta(-l) - \zeta(\lambda))$$

$$+ e^{l(-\nu + ip_1)} \frac{\sigma(l - \lambda)\sigma(\tau)}{\sigma(l)\sigma(\lambda)}$$

$$\times (x^{-1} + \zeta(l - \lambda) - \zeta(l) + \zeta(\tau)) + S(l).$$

On the other hand, this function can be written in the form

$$G(l, x) = G_0(l) e^{rx} \frac{\sigma_1(x + \mu)}{\sigma_1(x)},$$

where σ_1 is the Weierstrass function with quasiperiods $(1, \omega_2)$ and $G_0(l)$ is a constant factor. The parameters r and μ can be found from the quasi-periodicity conditions

$$r(\tau, p_1) = (2\pi i)^{-1} (\eta_{12}(\nu - ip_1) - (\tau - \lambda)\eta_2 \eta_{11},$$

$$\mu(\tau, p_1) = (2\pi i)^{-1} (-\omega_2(\nu - ip_1) + (\tau - \lambda)\eta_2),$$

where $\eta_{11} = 2\zeta_1(1/2)$ and $\eta_{12} = 2\zeta_1(\omega_2/2)$. Comparing the decompositions of both forms near $x = 0$, one finds the sum $S(l)$ explicitly:

$$S(l) = -\frac{\sigma(-l+\tau)}{\sigma(-l)}(\zeta_1(\mu) + r - \zeta(-l+\tau) + \zeta(-l) + \zeta(\lambda))$$

$$+\frac{\sigma(l-\lambda)\sigma(\tau)}{\sigma(l)\sigma(\lambda)}(\zeta_1(\mu) + r - \zeta(l-\lambda+\zeta(l) - \zeta(\tau))e^{l(-\nu+ip_1)}.$$

$$(3.160)$$

The explicit form of $F(\tau, p_1, p_2, j, k)$ now reads with the use of Equations (3.159) and (3.160) as follows:

$$F(\tau, p_1, p_2, j, k) = -\frac{\sigma(j-k+\tau)}{\sigma(j-k)}\varphi_0 t_0 e^{i(p_1 j + p_2 k)}$$

$$\times(\zeta_1(\mu) + r - \zeta(j-k+\tau) + \zeta(j-k) + \zeta(\lambda))$$

$$+ \varphi_0 \sigma(\tau)t(j-k)e^{i(p_1+p_2)k}$$

$$\times(\zeta_1(\mu) + r + \zeta(j-k+\lambda) - \zeta(j-k) - \zeta(\tau)).$$

Now, taking explicit summation of all four F' with different arguments in Equation (3.157), one finds its compact form

$$-\phi_{jk}t_0[\zeta_1(\mu) + \zeta_1(\tilde{\mu}) + r + \tilde{r} + 2\zeta(\lambda)] + \varphi_0\sigma(\tau)[t(j-k)e^{i(p_1+p_2)k}$$

$$+ t(k-j)e^{i(p_1+p_2)j}] \times [\zeta_1(\mu - \zeta_1(\tilde{\mu}) + r - \tilde{r} - 2\zeta(\tau)]$$

$$= E\phi_{jk} - g_0(t(j-k)e^{i(p_1+p_2)k} + t(k-j)e^{i(p_1+p_2)j},$$

where $\tilde{r} = r(-\tau, p_2)$, $\tilde{\mu} = \mu(-\tau, p_2)$. Comparing the coefficients in both sides, one obtains two equations, as follows:

$$E = -t_0[\zeta_1(\mu) + \zeta_1(\tilde{\mu}) + r + \tilde{r} + 2\zeta(\lambda)],$$

$$g_0 = -\varphi_0\sigma(\tau)[\zeta_1(\mu) - \zeta_1(\tilde{\mu}) + r - \tilde{r} - 2\zeta(\tau)].$$

These equations define E and g_0/φ_0 in terms of p_1, p_2 and τ. Plugging them into (3.156) results in the equation for determining the phase shift τ. Together with the relations between τ, p_1 and p_2 (3.154), this equation allows one to determine all the parameters of Ansatz (3.151) for two-electron wave functions. It would be of interest to investigate the question of the completeness of this solution.

To summarize, the presence of scalar currents commuting with Hamiltonian is the first evidence of the integrability for the Hubbard models with the hopping (3.146) presenting the "square root" for the elliptic exchange in Heisenberg–Dirac–van Vleck chains. It is possible to find the corresponding two-fermion function analytically [93]. However, it is seen also that the construction of higher scalar currents is an extremely hard problem and many-fermion wave functions could also be cumbersome and complicated. Till now, nothing is known about ground-state wave function of the simplest trigonometric Gebhard–Ruckenstein model: it is neither of Jastrow-type, like the Haldane–Shastry model, nor of Bethe Ansatz form, as in the case of the hopping (6.3).

3.7. Concluding Remarks

The main known facts about the integrable Heisenberg–Dirac–Van Vleck chains with variable range exchange and related Hubbard models were described. Many questions in their theory are still open.

As concerns the integrability of these models, understanding it from the Yang–Baxter viewpoint is highly desirable. For the spin chains, it is quite probable that the corresponding R matrix is the same as in [74]. The problem of mutual commutativity of the set of operators (2.6) might be solved in this way. Nothing is known for the integrability of elliptic Hubbard chains except of the simplest conserved current (3.146) and the two-fermion wave function.

The model with hyperbolic exchange on infinite line should have rich variety of multimagnon bound states which are given by solutions to transcendental equations $1 - (2\kappa)^{-1}[f(p_j) - f(p_k)] = 0$ as it follows from (3.33). It would be of interest to find a more simple way of constructing eigenfunctions of the Calogero–Moser Hamiltonian with inverse square hyperbolic particle interaction.

The exact equations of Bethe Ansatz type for the case of periodic boundary conditions are too complicated at the present stage of finding solutions to the quantum elliptic Calogero–Moser equation at $l = 1$. One cannot exclude the possibility of discovering their more simple form which would be of use to verify the hypothesis of asymptotic Bethe Ansatz in the thermodynamic limit. The construction described above does not allow either to do that or to establish the correspondence with the trigonometric Haldane–Shastry model.

In the models on inhomogeneous lattices, the main problem also consists in finding the proof of integrability for the most general potential of the external field (3.125). The simple formula for the spectrum for the case of the Morse potential, which comprises rational and Haldane–Shastry models, still waits analytical confirmation. If one would find the explicit form of the unitary transformation of the basic Hamiltonian to its simple effective form (3.129), a lot of results about various correlation functions would be obtained for the Haldane–Shastry chain.

The only known results about the spectra of the related Hubbard models are given by the original work of Gebhard and Ruckenstein [86]. The trigonometric and hyperbolic versions both have the $sl_2 \otimes sl_2$ Yangian symmetry and scalar integrals of motion (3.146). The most challenging problem is to prove the integrability and find the Bethe Ansatz-like formulas for the spectrum of the most general Hubbard model with elliptic hopping (3.149). Its solution could clarify the algebraic nature of the integrability of all the models under present consideration.

Chapter 4

Integrable Systems of Particles with Spin

4.1. Systems of Particles Interacting via Rational and Trigonometric Potentials

In view of the results of the previous chapter, it is natural to ask whether there exist integrable quantum systems of particles carrying spins. It turned out that it is possible. Indeed, Ha and Haldane [95] found the following generalization of the Sutherland Hamiltonian:

$$H = H^0 + H^1 + H^2,$$

$$H^0 = -\frac{1}{2} \sum_{i=1}^{N} \frac{\partial^2}{\partial x_j^2},$$

$$H^1 = \lambda^2 \sum_{i<j}^{N} \frac{1}{d(x_i - x_j)^2}, \qquad (4.1)$$

$$H^2 = \lambda \sum_{i<j}^{N} \frac{P_{ij}^\sigma}{d(x_i - x_j)^2},$$

where λ is coupling constant ($\lambda \in \mathbf{R}$), P_{ij}^σ is an operator that exchanges particle spins at x_i and x_j and $d(x)$ might be either distance between particles or chord distance on the circle with circumference L, $d(x) = (L/\pi) \sin(\pi x / L)$. Following Ha and Haldane [95], we shall consider further only the last case. If all particles have the same spin, the model reduces to the system of spinless particles and its solution was described already by Sutherland [105, 106]. It is connected to the rational case by the identity

$$\sum_{n=-\infty}^{+\infty} \frac{1}{(x+nL)^2} = \frac{1}{d(x)^2}.$$

The trial wave eigenfunction of the Hamiltonian (4.1) can be written as

$$\Psi(\{z\sigma\}) = \Psi_0 \prod_{k=1}^{N} z_k^{J_{\sigma_k}}, \qquad (4.2)$$

where

$$\Psi_0 = \prod_{n>m}^{N} \phi_{nm},$$

$$\phi_{nm} = |z_n - z_m|^{\lambda-x}(z_n - z_m)^{x+\delta_{\sigma_n\sigma_m}}$$

$$\times \exp\left[i\frac{1}{2}\pi \operatorname{sign}(\sigma_n - \sigma_m)\right]. \qquad (4.3)$$

Here, $x = 1$ for bosons and $x = 0$ for fermions, δ is the Kronecker delta function and $z_n = \exp(2\pi x_n/L)$, σ_n is the ordered spin index and J_σ is the global current of particles with σ spin. We take J_σ to be an integer and will discuss the restriction on the allowed values of J_σ later. Let us note also that the wave function with $\lambda = x = 0$ is the Slater determinant that corresponds to the states of free $SU(n)$ fermions.

The symmetry of the wave function with respect to the exchange of particles is given as

$$\Psi(\ldots, z_i\sigma_i, \ldots, z_j\sigma_j, \ldots) = (-1)^{x+1}\Psi(\ldots, z_j\sigma_j, \ldots, z_i\sigma_i, \ldots). \quad (4.4)$$

The total Hamiltonian (4.1) can be considered as a sum of kinetic, potential and exchange Hamiltonian, respectively. Each operator acting on the wave function (4.3) gives two types of terms, "wanted" and "unwanted". "Wanted" terms are defined to be the terms that depend only on the global variables such as the total number of particles, J_σ, etc. "Unwanted" terms explicitly depend on the local variables like z_i. Since the eigenenergy should depend only on the global variables, the "unwanted" terms arising in acting on H^0, H^1 and H^2 should cancel or combine to give "wanted" terms.

Let us first examine H^0 acting on the wave function. For this purpose, let us define the following derivatives:

$$\varphi_{ij} = \frac{\partial_{z_i}\phi_{ij}}{\phi_{ij}} = -\frac{\lambda}{z_i - z_j} - \frac{\lambda - x}{2z_j} - \frac{\delta_{\sigma_i\sigma_j}}{z_i - z_j}, \qquad (4.5)$$

$$\xi_{ij} = \partial_{z_j}\varphi_{ij}, \qquad (4.6)$$

$$\eta_j^{(1)} = z_j\frac{\partial_{z_j}\Psi_0}{\Psi_0} = z_i\sum_{i\neq j}\varphi_{ij}, \qquad (4.7)$$

$$\eta_j^{(2)} = z_j^2\frac{\partial_{z_j}^2\Psi_0}{\Psi_0} = z_j^2\sum_{i\neq j}\xi_{ij} + (\eta_j^{(1)})^2. \qquad (4.8)$$

After defining these derivatives, it is possible to write the action of H^0 to the wave function (4.3) as

$$H^0\Phi = 2(\pi/L)^2\sum_j[\eta_j^{(2)} + (2J_{\sigma_j} + 1)\eta_j(1) + J_{\sigma_j}^2]\Phi.$$

By performing some simple calculations, one can show that H^0 acting on the wave function gives "wanted" (W_K) and "unwanted" (U_K) terms,

$$H^0\Phi = 2\left(\frac{\pi}{L}\right)^2(W_K + U_K)\Phi, \qquad (4.9)$$

$$W_K = \frac{\lambda}{12}N(N^2 - 1) + \frac{1}{2}(\lambda + 1)\sum_\sigma M_\sigma(M_\sigma - 1)$$

$$+ \frac{1}{3}\sum_\sigma M_\sigma(M_\sigma - 1)(M_\sigma - 2)$$

$$- \sum_\sigma M_\sigma J_\sigma(\lambda N - M_\sigma - J_\sigma - \lambda + 1) + E(x), \qquad (4.10)$$

$$U_K = \lambda\sum_{i\neq j}\frac{z_i z_j}{(z_i - z_i)^2}(\lambda - 1 + 2\delta_{\sigma_i\sigma_j}) + \lambda\sum_{i\neq j}\frac{z_i J_{\sigma_i} - z_j J_{\sigma_j}}{z_i - z_j}$$

$$+ \lambda\sum_{i\neq j\neq k}\frac{z_i(z_i + z_k)}{(z_i - z_j)(z_i - z_k)}\delta_{\sigma_i\sigma_j}. \qquad (4.11)$$

where $E(x) = x^2 N(N-1)^2/4 + (x/2)(N-1)\sum_\sigma M_\sigma(M_\sigma - 1) + x(N-1)\sum_\sigma M_\sigma J_\sigma$, M_σ is the total number of σ particles in the system. One can now anticipate that U_K is canceled or combined with the "unwanted" terms from H^1 and H^2 to give local variable independent terms.

H^1 acts on Φ simply as $\lambda^2 \sum_{i<j} 1/d(x_i - x_j)^2$. The action of H^2 on the wave function is much more complicated. One needs to evaluate the expression

$$P_M = \frac{1}{2\Phi} \sum_{i>j} \frac{P_{ij}^\sigma \Phi}{d(x_i - x_j)^2}$$

$$= -\frac{1}{\Phi(\{z\sigma\})} \sum_{i \neq j} \frac{z_i z_j}{(z_i - z_j)^2} \Phi(\{z\sigma'\}), \qquad (4.12)$$

where $\{z\sigma'\}$ is a configuration with σ_i and σ_j exchanged with respect to $\{z\sigma\}$. Similarly, one gets $\{z'\sigma\}$ equal to $\{z\sigma\}$ with z_i and z_j exchanged. Using the identity $\Phi(\{z\sigma\}) = (-1)^{x+1}\Phi(\{z'\sigma'\})$, one rewrites the exchange operation as

$$P_M = (-1)^{x+1} \sum_{i \neq j} \frac{z_i z_j}{(z_i - z_j)^2} \frac{\Phi(z\sigma')}{\Phi(\{z'\sigma'\})}$$

$$= \sum_{i \neq j} \frac{z_i z_j}{(z_i - z_j)^2} (-1)^{\delta_{\sigma_i \sigma_j}} \left(\frac{z_i}{z_j}\right)^{J_{\sigma_j} - J_{\sigma_i}} \prod_{k \neq ij} \left(\frac{z_k - z_i}{z_k - z_j}\right)^{\delta_{\sigma_j \sigma_k} - \delta_{\sigma_i \sigma_k}}.$$

$$(4.13)$$

The action of P_M is the same for both boson and fermion cases and is independent of the interaction parameter λ.

Let us use the following identity:

$$\left(\frac{z}{z'}\right)^n = \sum_{q=0}^{|n|} \frac{|n|!}{(|n| - q)! q!} \left[\left(\frac{z - z'}{z'}\right)^q \theta(n) + \left(\frac{z' - z}{z}\right)^q \theta(-n)\right],$$

$$(4.14)$$

to rewrite Equation (4.13) as

$$P_M = P_0 + \sum_{\sigma, \sigma'} \sum_{q=1}^{|J_{\sigma'} - J_\sigma|} P_q^{\sigma \sigma'}, \qquad (4.15)$$

where θ in (4.14) is step function with $\theta(0) = 1/2$, and P_0 and $P_q^{\sigma\sigma'}$ are defined as

$$P_0 = \sum_{i \neq j} \frac{z_i z_j}{(z_i - z_j)^2} \left[-\delta_{\sigma_i \sigma_j} + (1 - \delta_{\sigma_i \sigma_j}) \right.$$

$$\left. \times \prod_{k \neq i,j} \left(\frac{z_k - z_i}{z_k - z_j} \right)^{\delta_{\sigma_j \sigma_k} - \delta_{\sigma_i \sigma_k}} \right], \tag{4.16}$$

$$P_q^{\sigma\sigma'} = 2 \sum_{i \neq j} \frac{z_i z_j}{(z_i - z_j)^2} \delta_{\sigma\sigma_i} \delta_{\sigma'\sigma_j} (1 - \delta_{\sigma\sigma'}) \prod_{k \neq i,j} \left(\frac{z_k - z_i}{z_k - z_j} \right)^{\delta_{\sigma_j \sigma_k} - \delta_{\sigma_i \sigma_k}}$$

$$\times \frac{(J_{\sigma'} - J_\sigma)!}{(J_{\sigma'} - J_\sigma - q)! q!} \theta(J_{\sigma'} - J_\sigma) \left(\frac{z_i - z_j}{z_j} \right)^q. \tag{4.17}$$

Note that it is useful to consider separately the terms with $q = 0, 1$ and the terms with $q > 1$. For the terms with $q = 0, 1$, one can introduce two sets of site indices $\{\alpha\}$ and $\{\beta\}$. The set $\{\alpha\}(\{\beta\})$ includes all the locations of particles with the spin $\sigma(\sigma')$. With the use of the identity

$$\left(\frac{z_k - z_i}{z_k - z_j} \right)^{\delta_{\sigma'\sigma_k} - \delta_{\sigma\sigma_k}} = 1 - \delta_{\sigma'\sigma_k} \frac{z_i - z_j}{z_k - z_j} + \delta_{\sigma\sigma_k} \frac{z_i - z_j}{z_k - z_i} \tag{4.18}$$

for $\sigma \neq \sigma'$, the products in P_M may be expanded and the typical terms in P_M can be simplified using two propositions.

Proposition 1. *Let $\{\alpha\}$ and $\{\beta\}$ be two sets of integers between 1 and N, and let $0 \leq q \leq 1$ and*

$$\Delta = (1 - \delta_{\sigma_i \sigma'}) \delta_{\sigma\sigma_i} \delta_{\sigma\sigma_{\alpha_1}} \cdots \delta_{\sigma\sigma_{\alpha_n}} \delta_{\sigma'\sigma_j} \delta_{\sigma'\sigma_{\beta_1}}.$$

Let us denote also

$$\Delta_{MM'} = \sum_{n=1}^{M_\sigma - 1} \sum_{m=1-q}^{M_{\sigma'} - 1} \sum_{i \neq j} \sum_{(\alpha),(\beta)} \Delta \frac{(-1)^m}{n! m!}$$

$$\times \frac{z_i z_j^{1-q} (z_i - z_j)^{n+m-2+q}}{(z_{\alpha_1} - z_i) \cdots (z_{\alpha_n} - z_i)(z_{\beta_1} - z_j) \cdots (z_{\beta_m} - z_j)}.$$

Then,

$$\Delta_{MM'} = - \sum_{k=1}^{\min(M_\sigma, M_{\sigma'})} (M_\sigma - k)(M_{\text{sigma}'} - k)$$

for q = 0, and

$$\Delta_{MM'} = - \sum_{k=1}^{\min(M_\sigma, M_{\sigma'})} (M_\sigma - k)$$

for q = 1.

Proof of the Proposition 1. Note first that the indices $\{i, \alpha_1, \ldots, \alpha_n\}$ and $\{j, \beta_1, \ldots, \beta_m\}$ are dummy indices. Consequently, the sum will be invariant under any permutation of the indices. Now observe that Δ is invariant under any permutation of the indices except when there is interchange between two sets $\{i, \alpha_1, \ldots, \alpha_n\}$ and $\{j, \beta_1, \ldots, \beta_m\}$. On the other hand, the prefactor to Δ remains invariant with respect to the permutations of indices $\{\alpha\}$ and $\{\beta\}$. It is convenient to consider only the cyclic permutations which give $(n+1)(m+1)$ distinct prefactors. After summing the $(n+1)(m+1)$ factors, one obtains the following result:

$$P_{nm}^{\sigma\sigma'} = \sum_{i \neq j} \sum_{\{\alpha\},\{\beta\}} \frac{(-1)^m}{n!m!}$$

$$\times \frac{z_i z_j^{1-q}(z_i - z_j)^{n+m-2+q}}{(z_i - z_{\alpha_1}) \cdots (z_i - z_{\alpha_n})(z_j - z_{\beta_1}) \cdots (z_j - z_{\beta_m})} \Delta$$

$$= \sum_{i \neq j} \sum_{\{\alpha\},\{\beta\}} \sum_{k,l} \frac{(-1)^m}{(n+1)!(m+1)!}$$

$$\times \frac{z_{\gamma_k^\alpha} z_{\gamma_l^\beta}^{1-q}(z_{\gamma_k^\alpha} - z_{\gamma_l^\beta})^{n+m-2+q}}{(z_{\gamma_k^\alpha} - z_{\alpha_1}) \cdots (z_{\gamma_k^\alpha} - z_i) \cdots (z_{\gamma_k^\alpha} - z_{\alpha_n})}$$

$$\times \frac{\Delta}{(z_{\gamma_l^\beta} - z_{\beta_1}) \cdots (z_{\gamma_l^\beta} - z_j) \cdots (z_{\gamma_l^\beta} - z_{\beta_m})},$$

where $\{\gamma^\alpha\} = \{i, \alpha_1, \ldots, \alpha_n\}$ and $\{\gamma^\beta\} = \{j, \beta_1, \ldots, \beta_m\}$. After some algebra, the sum transforms as

$$P_{nm}^{\sigma\sigma'} = \frac{(-1)^{n+m-1+q}}{(n+1)!(m+1)!} \frac{(n+m-2+q)!}{(m-1+q)!(n-1)!} \sum_{i\neq j} \sum_{\{\alpha\},\{\beta\}} \Delta$$

$$= (-1)^{n+m-1+q} \frac{(n+m-2+q)!}{(m-1+q)!(n-1)!} \frac{M_\sigma!}{(M_\sigma - n - 1)!(n+1)!}$$

$$\times \frac{M_{\sigma'}!}{(M_{\sigma'} - m - 1)!(m+1)!}$$

for $0 \leq q \leq 1$. Here, evaluating $\sum_{i\neq j} \sum_{\{\alpha\},\{\beta\}} \Delta$ is the same as calculating the total numbers of ways of putting $n+1$ out of M_σ blue balls in a box and $m+1$ out of $M_{\sigma'}$ red balls in another box. The sum, therefore, is equal to $M_\sigma(M_\sigma - 1)\cdots(M_\sigma - n)M_{\sigma'}(M_{\sigma'} - 1)\cdots(M_{\sigma'} - m)$.

It is convenient to use the following identities:

$$\frac{(n-m-2+q)!}{(m-1+q)!(n-1)!} = \sum_{s=0} \frac{(n-1)!}{(n-1-s)!s!} \frac{(m-1+q)!}{(m-1+q-s)!s!},$$

$$(4.19)$$

$$S_M = \sum_{m=1-q}^{M_\sigma - 1} (-1)^m \frac{M_\sigma!}{(M_\sigma - m - 1)!(m+1)!} \frac{(m+1)!}{(m+1-k)!k!}$$

$$(4.20)$$

$= \text{(for } q = 0\text{)} \ -M_\sigma + 1 \text{ if } k = 0, \ -M_\sigma \text{ if } k = 1 \text{ and } 0 \text{ if } 2 \leq k < M_\sigma,$

$$S_M = 1 \text{ if } k = 0, 0 \text{ if } 1 \leq k < M_\sigma \quad \text{for } q = 1.$$

Using (4.19) one can sum the terms depending on n and m in $P_{nm}^{\sigma\sigma'}$ separately. Using (4.20) the following relation can be obtained:

$$P^{\sigma\sigma'} = - \sum_{k=1}^{\min(M_\sigma, M_{\sigma'})} (M_\sigma - k)(M_{\sigma'} - k) \quad \text{for } q = 0,$$

$$P^{\sigma\sigma'} = - \sum_{k=1}^{\min(M_\sigma, M_{\sigma'})} (M_\sigma - k) \quad \text{for } q = 1.$$

Proposition 1 is proved. $\qquad\qquad\qquad\qquad\qquad\qquad\square$

Proposition 2. *For $q \geq 2$ and $M_\sigma \geq M_{\sigma'}$ the following equality takes place. Let*

$$P = \sum_{i \neq j} \frac{z_i}{z_j} \left(\frac{z_i - z_j}{z_j} \right)^{q-2} (1 - \delta_{\sigma\sigma'}) \delta_{\sigma\sigma_i} \delta_{\sigma'\sigma_j} \prod_{k \neq i,j} \left(\frac{z_k - z_i}{z_k - z_j} \right)^{\delta_{\sigma'\sigma_k} - \delta_{\sigma\sigma_k}}.$$

Then $P = M_{\sigma'}$ for $q = 2$ and $P = 0$ for $2 < q \leq M_\sigma - M_{\sigma'} + 1$.

Proof. Let us replace the product over k in P with two new sets of integer indices, $\{\alpha\}$ and $\{\beta\}$. Then

$$P = \sum_{i \neq j} \sum_{\{\alpha\},\{\beta\}} \frac{z_i}{z_j(M_\sigma - 1)!(M_{\sigma'} - 1)!} \left(\frac{z_i - z_j}{z_j} \right)^{q-2}$$

$$\times \frac{(z_{\alpha_1} - z_j)(z_{\alpha_2} - z_j) \cdots (z_{\alpha_{M_\sigma-1}} - z_j)}{(z_{\alpha_1} - z_i)(z_{\alpha_2} - z_i) \cdots (z_{\alpha_{M_\sigma-1}} - z_i)}$$

$$\times \left[\sum_{\tau=0}^{M_{\sigma'}-1} (z_i - z_j)^\tau f_\tau(\{z_\beta\}) \right] \Delta,$$

where Δ is the same as in Proposition 1, f_τ is some function of $\{z_\beta\}$ with $f_0 = 1$ and the factor in brackets is obtained by writing the terms like $(z_{\beta_1} - z_i)$ as $[1 - (z_i - z_j)/(z_{\beta_1} - z_j)]$ and by multiplying out all the factors that depend on $\{\beta\}$.

It is possible, as in proof of Proposition 1, to rewrite the expression by summing over the factors obtained by the cyclic permutations of the indices $\{i, \alpha_1, \ldots, \alpha_{M_\sigma-1}\}$ and obtain

$$P = \sum_{i \neq j} \sum_{\{\alpha\},\{\beta\}} \sum_{\tau=0}^{M_{\alpha'}-1} \frac{(-1)^{M_\sigma}}{M_\sigma!(M_{\sigma'}-1)!}(z_i - z_j)$$

$$\times (z_{\alpha_1} - z_j) \cdots (z_{\alpha_{M_\sigma}-1} - z_j)f_\tau \frac{W_\alpha^{M_\sigma}}{D_\alpha},$$

where $W_\alpha^{M_\sigma}$ is given by the Vandermonde determinant whose last row is modified to $z_{\gamma_1}(z_{\gamma_1} - z_j)^{q+\tau-3}/z_j^{q-1}$, $\gamma_l = \{i, \alpha_1, \ldots, \alpha_{M_\sigma-1}\}$ and $D_\alpha = \prod_{k<k'}(z_{\gamma_k^\alpha} - z_{\gamma_{k'}^\alpha})$. Now it is straightforward to show that

$$W_\alpha^{M_\alpha} = \frac{(-1)^{M_\sigma}D_\alpha}{(z_i - z_i)(z_{\alpha_1} - z_j) \cdots (z_{\alpha_{M_{\alpha}-1}} - z_j)} \quad \text{if } q = 2 \quad \text{and } \tau = 0,$$

$$W_\alpha^{M_\alpha} = 0 \quad \text{if } 0 \leq q + \tau - 3 \leq M_\sigma - 3,$$

$$\sum_{i \neq j} \sum_{\{\alpha\},\{\beta\}} \Delta = M_\sigma!M_{\sigma'}!.$$

The only non-zero contribution to P is given by the terms with $q = 2$ and $\tau = 0$. Since $0 \leq \tau \leq M_{\sigma'} - 1$, the sufficient condition for $W_\alpha^{M_\sigma} = 0$ is $3 \leq q \leq M_\sigma - M_{\sigma'} + 1$. Therefore, for $M_\sigma \geq M_{\sigma'}$

$$P = M_{\sigma'} \quad \text{for } q = 2,$$

$$P = 0 \quad \text{for } 3 \leq q \leq M_\sigma - M_{\sigma'} + 1.$$

Proposition 2 is proved. □

Now it is possible, after reorganizing some terms in (4.12), to obtain the following results:

$$P_M = W_P + U_P,$$

where

$$W_P = -\frac{1}{3}\sum_\sigma M_\sigma(M_\sigma - 1)(M_\sigma - 2)$$

$$-\frac{1}{3}\sum_{\sigma<\sigma'} M_{\sigma'}(M_{\sigma'} - 1)(3M_\sigma - M_{\sigma'} - 1)$$

$$-\sum_{\sigma<\sigma'} M_{\sigma'}(M_\sigma - M_{\sigma'})(J_{\sigma'} - J_\sigma) + \sum_{\sigma<\sigma'} M_{\sigma'}(J_{\sigma'} - J_\sigma)^2,$$

$$U_p = \sum_{i \neq j} \frac{z_i z_j}{(z_i - z_j)^2}(1 - 2\delta_{\sigma_i \sigma_j}) + \sum_{i \neq j} \frac{z_i J_{\sigma_j} - z_j J_{\sigma_i}}{z_i - z_j}$$

$$- 2 \sum_{(i \neq j \neq k)} \frac{z_i z_j}{(z_i - z_j)(z_i - z_k)}\delta_{\sigma_i \sigma_k}.$$

Because of symmetry, one can choose $M_1 \geq M_2 \geq \cdots \geq M_n$ and $0 \leq J_{\sigma'} - J_\sigma \leq M_\sigma - M_{\sigma'} + 1$ for $\sigma' > \sigma$ without loss of generality. And, one can easily check that U_K, U_P, and the potential-energy term combine to give local-variable-independent terms.

Finally, the eigenenergies of the model are given by $E = 2\pi^2/L^2(E_1 + E_2)$, where E_1 and E_2 are energies due to one- and two-spin interactions,

$$E_1 = \frac{\lambda^2}{12}N(N^2 - 1) + \frac{1}{2}\lambda N \sum_\sigma M_\sigma(M_\sigma - 1)$$

$$+ \frac{1}{6}(1 - \lambda) \sum_\sigma M_\sigma(M_\sigma - 1)(2M_\sigma - 1)$$

$$+ \sum_\sigma J_\sigma M_\sigma(M_\sigma + J_\sigma - 1),$$

$$E_2 = -\frac{1}{3}\lambda \sum_{\sigma < \sigma'} M_{\sigma'}(M_{\sigma'} - 1)(3M_\sigma - M_{\sigma'} - 1)$$

$$+ \sum_{\sigma < \sigma'} \lambda[M_{\sigma'}(M_{\sigma'} - M_\sigma)(J_{\sigma'} - J_\sigma) + M_{\sigma'}(J_{\sigma'} - J_\sigma)^2].$$

The ground state is given by the following two conditions: (i) $M_\sigma = M_{\sigma'} = M$ and (ii) $J_{\sigma'} = J_\sigma = J = -(M - 1)/2, -(N + M - 1)/2$ for bosons and fermions, respectively. Thus, the model of Ha and Haldane is exactly solvable. Two generalizations of it will be considered in detail in the next two sections.

4.2. A Special Set of Eigenvectors for the Hyperbolic Sutherland Systems with Spin in the Morse Potential

The extension of the family of solvable models of the CS type has been studied. Within this approach, the most general models which have been constructed so far include the pairwise potential of short-range form $[a^{-1}\sinh(ax)]^{-2}$ and the interaction with three-parametric potential $W(x) = A_1 \cosh(4ax + b_1) + A_2 \cosh(2ax + b_2)$ [9]. The many-particle Hamiltonian is written as

$$H = \sum_{j}^{N} \left[-\frac{1}{2} \left(\frac{\partial}{\partial x_j} \right)^2 + W(x_j) \right] + \sum_{j<k}^{N} \frac{\lambda(\lambda+1)a^2}{\sinh^2 a(x_j - x_k)}. \qquad (4.21)$$

However, for generic values of parameters the simplicity of the eigenstates is lost. The only case in which it is restored at least partially corresponds to the limit $b_{1,2} \to +\infty$ upon rescaling A_α with $\exp(-b_\alpha)$, which results in the one-parametric Morse potential

$$W(x) = 2\tau^2 a^2 (\exp(2ax) - 1)^2. \qquad (4.22)$$

The dynamics of the systems described by (4.21) and (4.22) (hereafter to be called SM) has been found [10] to be more complicated.

Further progress has been made a few years ago by Ha and Haldane [95]. They proposed the generalization of the CS model to the particles with internal degrees of freedom ($SU(n)$ spins) and exchange spin interaction of the form $\lambda(\lambda + P_{jk})V(x_j - x_k)$, where the operator P_{jk} transposes the spins of particles labeled by indices j and k,

$$P_{jk} = \sum_{p,q=1}^{n} e_j^{pq} e_k^{qp}, \qquad (4.23)$$

$\{e_j^{pq}\}$ are the elementary spin operators obeying the commutation relations

$$[e_j^{pq}, e_k^{rs}] = \delta_{jk}(\delta^{rq} e_j^{ps} - \delta^{ps} e_j^{rq}). \qquad (4.24)$$

These new models and corresponding lattice spin chains with clear physical interpretation, which arise upon "freezing" dynamical degrees of freedom in the limit of large coupling constant λ in the Hamiltonians like (4.25), have been investigated in great detail [109–119]. As concerns their spectral properties, the common feature is the presence of huge degeneracies which have been successfully classified by identifying the internal symmetry of Yangian type [111, 116]. As for more complicated systems of the SM type, the spectral problem has been studied only for the family of non-uniform spin lattices [118].

In view of the results of [110–119], a natural question is to consider whether there exists a solvable spin version of the dynamical SM systems described by the Hamiltonian

$$H_{SM}^{(s)} = \sum_{j}^{N} \left[-\frac{1}{2} \left(\frac{\partial}{\partial x_j} \right)^2 + 2\tau^2 a^2 (\exp(2ax_j) - 1)^2 \right]$$

$$+ \sum_{j<k}^{N} \frac{\lambda(\lambda + P_{jk})a^2}{\sinh^2 a(x_j - x_k)}. \tag{4.25}$$

The aim of the present section is to show that the answer is positive.

Now it is well-recognized that the possibility of the reduction of the spectral problem in quantum mechanics is tightly connected with the underlying symmetry of the systems with N degrees of freedom, which results in the existence of at least $N - 1$ operators commuting with the Hamiltonian. In what follows, it will be demonstrated that this way is appropriate for the treatment of the SM spin systems.

To simplify the notation, it is useful to introduce the variables $z_j = \exp(2ax_j)$ in which the Hamiltonian (4.25) acquires the more convenient form $H_{SM}^{(s)} = 4a^2 \mathcal{H}_{SM}^{(s)}$,

$$\mathcal{H}_{SM}^{(s)} = \sum_{j=1}^{N} \left[\frac{1}{2}\hat{p}_j^2 + w(z_j) \right] + \sum_{j<k}^{N} \frac{\lambda(\lambda + P_{jk})z_j z_k}{(z_j - z_k)^2}, \tag{4.26}$$

where

$$w(z) = \frac{\tau^2}{2}(z_j - 1)^2 \tag{4.27}$$

and $\hat{p}_j = -iz_j \partial/\partial z_j$. Hereafter it will be assumed that $\lambda > 0$.

The basic Lax relation for the operator (4.26) reads

$$\left[\mathcal{H}_{SM}^{(s)}, L \right] = [L, M]. \tag{4.28}$$

The invariants of the matrix L do not commute with (4.26). However, if one finds the pair (L, M) such that M obeys the condition

$$M\mathcal{I} = \mathcal{I}^\dagger M = 0 \tag{4.29}$$

with the one-column vector $\mathcal{I}$ with equal entries, the relation (4.28) guarantees the existence of the integrals of motion in the form [98]

$$I_s = \mathcal{I}^\dagger L^s \mathcal{I}. \tag{4.30}$$

To construct the Lax pair for the SM spin systems, let us consider the following Ansatz [107]:

$$L = \begin{pmatrix} L_0 & \psi + \rho \\ -\psi - \rho & -L_0 \end{pmatrix} \quad M = \begin{pmatrix} M_0 + m & \phi \\ \phi & M_0 + m \end{pmatrix}, \tag{4.31}$$

where L_0 and M_0 constitute the standard Lax pair for the Sutherland N-particle system,

$$(L_0)_{jk} = -\frac{i}{2}\lambda f_{jk} S_{jk}(1 - \delta_{jk}) + \hat{p}_j \delta_{jk},$$

$$(M_0)_{jk} = i(1 - \delta_{jk})\lambda h_{jk} S_{jk} - i\lambda \delta_{jk} \sum_{s \neq j}^{N} h_{js} S_{js}, \tag{4.32}$$

$$f_{jk} = \frac{z_j + z_k}{z_j - z_k}, \quad h_{jk} = \frac{z_j z_k}{(z_j - z_k)^2}$$

and ψ, ϕ, ρ and m are $(N \times N)$ matrices with entries

$$\begin{aligned} (\psi)_{jk} &= \xi(z_j)\delta_{jk}, \quad (\phi)_{jk} = \varphi(z_j)\delta_{jk}, \\ (m)_{jk} &= \mu(z_j)\delta_{jk}, \quad (\rho)_{jk} = cP_{jk}(1 - \delta_{jk}). \end{aligned} \tag{4.33}$$

One finds that the Lax relation (4.28) is equivalent to the overdetermined set of functional equations

$$iz\frac{d}{dz}(w(z) + \mu(z)) = 2\xi(z)\varphi(z), \quad \varphi(z) = -\frac{i}{2}z\frac{d\xi}{dz}, \tag{4.34}$$

$$-\frac{i\lambda}{2}f_{jk}[\mu(z_k) - \mu(z_j)] + c[\varphi(z_j) + \varphi(z_k)] = 0, \tag{4.35}$$

$$-\frac{i\lambda}{2}f_{jk}[\varphi(z_j) + \varphi(z_k)] + \lambda h_{jk}[\xi(z_j) - \xi(z_k)] = c[\mu(z_j) - \mu(z_k)]. \tag{4.36}$$

Starting from the most restrictive equations (4.35) and (4.36), one obtains the general solution of this set in the form

$$\mu(z) = \mu_1 z + \mu_2 z^{-1}, \quad \varphi(z) = \epsilon(\mu_1 z - \mu_2 z^{-1}),$$

$$\xi(z) = 2i\epsilon(\mu_1 z + \mu_2 z^{-1} + \gamma), \tag{4.37}$$

$$w(z) = 2[\mu_1^2 z^2 + \mu_2^2 z^{-2} + (2\gamma - 1/2)(\mu_1 z + \mu_2 z^{-1})],$$

$$2c = -i\epsilon\lambda, \quad \epsilon = \pm 1. \tag{4.38}$$

One can see from (4.33) that the matrix M (4.31) obeys the condition (4.29) if and only if $\mu(z) + \varphi(z) = 0$. From (4.37) and (4.38), it results that $w(z)$ acquires the form (4.27) and the entries of the matrices L and M are written as

$$(\psi)_{jk} = -i[\tau(z_j - 1) + 1/2]\delta_{jk}, \quad (\phi)_{jk} = -(m)_{jk} = -\tau z_j \delta_{jk}/2,$$

$$(\rho)_{jk} = i\lambda S_{jk}(1 - \delta_{jk})/2. \tag{4.39}$$

Thus, the set of integrals of motion is given by the formulae (4.30), (4.31), (4.32), (4.39) at even values of $s = 2l$, $1 \leq l \leq N$. It is easy to check that the Hamiltonian (4.26) is proportional to the first element I_2 of this set and all $\{I_{2l}\}$ are functionally independent.

The existence of the Lax pair (4.31) with the property (4.29) allows to construct another set of operators commuting with the Hamiltonian (4.26) similar to that found in [98]. Let $E^{ab} = \text{diag}(e_1^{ab}, \ldots, e_N^{ab}, e_1^{ab}, \ldots, e_N^{ab})$ be $(2N \times 2N)$ matrix with entries obeying the relations (4.24). From (4.31) and (4.32) one can see that (4.28) still holds if L is replaced by E^{pq}. Thus, the corresponding set is given by

$$Q_l^{ab} = 2^{-(l+1)} \mathcal{I}^\dagger E^{ab} L^{2l} \mathcal{I}. \tag{4.40}$$

In what follows, the consideration will be restricted to the case of $s = 1/2$ fermions which is of most importance due to the possibility of the application of the results to the description of the phenomena in confined quasi-1D electron gas [115]. To find the ground-state energy of the systems described by the Hamiltonian (4.25), it is natural to use the trial wave function of Jastrow type [103, 105] as it has been done in [108, 110],

$$\Psi(z_1\sigma_1,\ldots,z_N\sigma_N) = C^{-1/2} \prod_{1\leq j<k\leq N} |z_j - z_k|^\lambda (z_j - z_k)^{\delta_{\sigma_j\sigma_k}}$$

$$\times \exp\frac{i\pi}{2}\mathrm{sgn}(\sigma_j - \sigma_k)$$

$$\times \prod_{l=1}^{N} z_l^{\rho+\Delta\sigma_l} \exp(-\tau z_l). \tag{4.41}$$

Here, $\sigma_j = 2s_j^{(3)}$, $s_j^{(3)} = \pm 1/2$ are the values of the spin projection of jth particle. The parameters ρ and Δ have to be chosen from the Schrödinger equation

$$\mathcal{H}_{SM}^{(s)}\Psi = [A_1 + \tau A_2(z,\sigma) + \lambda A_3(z,\sigma)]\Psi = E\Psi,$$

where

$$A_1 = \mathrm{const} = -Na^2\lambda^2(N-1)(2N-1)/3,$$

$$A_2(z,\sigma) = \left[\rho - \tau + (N-1)\left(\lambda + \frac{1}{2}\right)\right] \sum_{k=1}^{N} z_k$$

$$+ \left(\Delta + \frac{N_+ - N_-}{2}\right) \sum_{k=1}^{N} \sigma_k z_k, \tag{4.42}$$

$$A_3(z,\sigma) = \frac{\varepsilon}{2} \sum_{k\neq 1}^{N} \frac{\sigma_k z_k - \sigma_l z_l}{z_l - z_k} - \sum_{k\neq l\neq m}^{N} \frac{z_k^2 \delta_{\sigma_k\sigma_l}}{(z_k - z_l)(z_k - z_m)}$$

$$+ \frac{1}{2}\sum_{k\neq l} \frac{z_k z_l}{(z_k - z_l)^2}(1 - \delta_{\sigma_k\sigma_l})$$

$$\times \left[1 - \left(\frac{z_l}{z_k}\right)^{\varepsilon(\sigma_k-\sigma_l)} \prod_{j\neq k,l}^{N} \left(\frac{z_j - z_l}{z_j - z_k}\right)^{\delta_{\sigma_j\sigma_k}-\delta_{\sigma_j\sigma_l}}\right]. \tag{4.43}$$

The last contribution to the right-hand side of (4.43) comes from the spin exchange part of the Hamiltonian (4.26).

One finds that the terms linear in $\{z\}$ which are present in $A_2(z,\sigma)$ vanish if the parameters ρ and Δ obey the conditions

$$\rho = \tau - (\lambda + 1/2)(N-1), \quad \Delta = \frac{1}{2}(N_- - N_+). \tag{4.44}$$

The explicit dependence on spin variables in the formula (4.43) is removed if one considers separately the coordinates of the particles with $\sigma = \pm1$ and denotes $\{z_j\}_{\sigma=1} = \{p_\nu\}$, $\nu = 1, \ldots, N_+$, $\{z_j\}_{\sigma=-1} = \{q_\mu\}$, $\mu = 1, \ldots, N_-$,

$$A_3(z,\sigma) = \frac{N-1}{4}(N_+ - N_-)^2 - \frac{1}{3}[N_+(N_+ - 1)(N_+ - 2)$$

$$+ N_-(N_- - 1)(N_- - 2)]$$

$$+ \sum_{\nu=1}^{N_+}\sum_{\mu=1}^{N_-}\left\{\frac{(N_+ - N_-)(p_\nu + q_\mu)}{2(p_\nu - q_\mu)}\right.$$

$$-\sum_{l\neq\nu}^{N_+}\frac{p_\nu^2}{(p_\nu - p_l)(p_\nu - q_\mu)} - \sum_{m\neq\mu}^{N_-}\frac{q_\mu^2}{(q_\mu - q_m)(q_\mu - p_\nu)}\right\}$$

$$+ \sum_{\nu=1}^{N_+}\sum_{\mu=1}^{N_-}\frac{p_\nu q_\mu}{(p_\nu - q_\mu)^2}$$

$$\times \left[1 - \left(\frac{p_\nu}{q_\mu}\right)^{N_+-N_-}\prod_{l\neq\nu}^{N_+}\frac{p_l - q_\mu}{p_l - p_\nu}\prod_{m\neq\mu}^{N_-}\frac{q_m - p_\nu}{q_m - q_\mu}\right]. \tag{4.45}$$

This expression seems to be enormously cumbersome. However, it can be treated successfully by using the following trick. Consider the integral

$$I_{N_+,N_-} = \int_{\mathcal{C}}\int_{\mathcal{C}'}\frac{dzdz'}{(z-z')^4}z^{N_+-N_-+1}z'^{N_--N_++1}$$

$$\times \prod_{\nu=1}^{N_+}\frac{z'-p_\nu}{z-p_\nu}\prod_{\mu=1}^{N_-}\frac{z-q_\mu}{z'-q_\mu}. \tag{4.46}$$

Let us choose the contours $\mathcal{C}$, $\mathcal{C}'$ encircling $\{p_\nu\}$, $\{q_\mu\}$ as follows: $\mathcal{C}'$ lies within $\mathcal{C}$ at $N_+ - N_- \geq 0$, $\mathcal{C}$ lies within $\mathcal{C}'$ at $N_- - N_+ \geq 0$ and the inner contour does not encircle 0 in both cases. By expanding the outer contour, it is easy to see that $I_{N_+,N_-} = 0$. On the other hand, its evaluation yields the double and triple sums of the form (4.45). One finds that $A_3(z,\sigma)$ does not depend on $\{z,\sigma\}$. Thus, the Jastrow-type functions (4.41) are indeed the eigenfunctions of the Hamiltonian (4.25) provided that ρ and τ are given by (4.44) with eigenvalues

$$E_\Delta = a^2 \left\{ -\frac{N}{6} [2\lambda^2(N-1)(2N-1) + \lambda(4N^2 - 3N - 4) + N^2 - 1] \right.$$

$$+ \tau N[2\lambda(N-1) + N] + \frac{1}{3}\Delta(\Delta^2 - 1)$$

$$\left. + \Delta^2 \left[\tau - N\left(\lambda + \frac{1}{2}\right) + \frac{\lambda}{2} \right] \right\}. \tag{4.47}$$

The minimum of (4.47) corresponds to the value of $\Delta = 0$ for N even and $\Delta = 1$ for N odd. One can apply the usual argumentation to validate the conjecture about the ground state: it exists if the strength of the Morse potential obeys the relation $\tau - (\lambda + 1/2)(N - 1) - \Delta/2 > 0$ and is described by (4.41), (4.47) at minimal possible value of Δ.

Thus, it has been shown that some results obtained for the CS systems of particles with spin can be extended to the SM ones. The Lax formalism provides the rich family of the conserved operators which constitute the representation of the underlying infinite-dimensional algebra. However, for the SM systems there is no analogs of the operators of creation and annihilation type which have been found for the confined Calogero model [116, 119] and applied successfully to the algebraic construction of the eigenstates.

It has been demonstrated that for the SM systems of $s = 1/2$ fermions the use of the Jastrow-type form of the wave functions allows one to find the ground-state energy and some part of the discrete spectrum characterized by "spin" excitations. There should also be the excitations of the "spatial" type similar to those found for the spinless SM systems of bosons [105, 107]. However, it is not yet clear how to modify the Ansatz (4.41) so as to take them into consideration.

4.3. Toward the Proof of Complete Integrability of Quantum Elliptic Many-Particle Systems with Spin Degrees of Freedom

The problem which will be considered in this section has a long story. It began in 1975 when Calogero [96] proposed new classical integrable many-particle systems with pairwise interaction proportional to the elliptic Weierstrass $\wp$ function with two periods ω_1 and ω_2 (usually $\omega_1 \in \mathbf{R}$, $\omega_2 \in i\mathbf{R}$), and the Hamiltonian

$$H = \sum_{j=1}^{N} p_j^2 + \sum_{j \neq k}^{N} \lambda^2 h_{jk}, \quad \lambda \in \mathbf{R}, \quad h_{jk} = \wp(x_j - z_k), \qquad (4.48)$$

where N is the number of particles, $p_j = \frac{dx_j}{dt}$ are their momenta, and here and throughout this section the short notation will be used,

$$f_{jk} = f(x_j - x_k) \qquad (4.49)$$

for any function of particle coordinates $\{x_j\}$. In the pioneering paper [96], the Lax representation $\frac{dL}{dt} = [L, M]$ has been constructed for these systems, and this fact — on the classical level — leads directly to the existence of N integrals of motion as invariants of the matrix L. The complete integrability, i.e. vanishing of Poisson brackets of these integrals, was proved 8 years later by Olshanetsky and Perelomov [97]. The Lax matrices which have been proposed in [69] are of the form

$$L_{jk} = p_j \delta_{jk} + \lambda(1 - \delta_{jk}) f_{jk},$$

$$M_{jk} = -\lambda^2 \delta_{jk} \sum_{s=1}^{N} f_{js} f_{sj} + (1 - \delta_{jk}) \lambda f'_{jk}, \qquad (4.50)$$

where $j, k = 1, \ldots, N$ (N being the number of particles), prime means differentiation of the function with respect to the argument and δ_{jk} is the Kronecker symbol. The function f is any solution of the functional equation [96]

$$f_{js} f'_{sl} - f'_{js} f_{sl} = f_{jl}(h_{js} - h_{sl}), \quad h_{js} = -f_{js} f_{sj}. \qquad (4.51)$$

In [96], some particular solutions to (4.51) were found in terms of combinations of the elliptic Jacobi functions, but the general solution to this equation had been identified only five years later in [53],

$$f_{jk} = \frac{\sigma(x_j - x_k + \alpha)}{\sigma(x_j - x_k)\sigma(\alpha)}, \tag{4.52}$$

where $\sigma(x)$ is the Weierstrass sigma function related to $\wp(x)$ as $\sigma'(x) = \sigma(x)\zeta(x)$, $\zeta'(x) = -\wp(x)$, and $\alpha \in \mathbf{C}$ is the spectral parameter which does not influence the dynamics,

$$h_{js} = -f_{js}f_{sj} = \wp(x_j - x_s) - \wp(\alpha). \tag{4.53}$$

The integrability of quantum mechanical version of (4.48) has been proved in [54], also based on the form of the Lax matrices L and M, with $p_j = -i\frac{\partial}{\partial x_j}$ now. Much later, it was noticed that the quantum Lax relation $[H, L] = [L, M]$ is satisfied if

$$H = \sum_{J \neq k}^{N} \wp_N(j - k)P_{jk}, \tag{4.54}$$

$$L_{jk} = (1 - \delta_{jk})f_{jk}P_{jk},$$

$$M_{jk} = -\delta_{jk}\sum_{s \neq j}^{N} f_{js}f_{sj}P_{js} + (1 - \delta_{jk})f'_{jk}P_{jk}, \tag{4.55}$$

where $\wp_N(j)$ is the Weierstrass $\wp$ function with real period N (the second period, ω_2, remains arbitrary), f_{jk} is constructed from sigma functions with quasiperiods N and ω_2 as in (4.52), and $\{P_{jk}\}$ form any representation of the symmetric group S_N (for physical applications, the most important representation is given by Pauli matrices),

$$P_{jk} = \frac{1}{2}(1 + \vec{\sigma}_j\vec{\sigma}_k).$$

In this case, (4.54) becomes the Hamiltonian of the homogeneous quantum spin chain with variable range exchange. Its effective range is controlled by the parameter $|\omega_2|$: as $|\omega_2| \to 0$, (4.54) is the Hamiltonian of the Heisenberg chain with nearest-neighbor exchange; as $|\omega_2| \to \infty$, one gets long-ranged Haldane–Shastry spin chain. But in the elliptic case the existence of Lax pair gives nothing for proving integrability! In fact, one cannot find even the Hamiltonian among the traces of various degrees of the L matrix. Nevertheless (and surprisingly enough!), it turned out that integrals of motion can be constructed by using the function f_{jk} (the same as in the definition of L matrix (4.55). Guided rather by intuition, two of them have been constructed in [58] as

$$I_{1,2} = \sum_{j \neq k \neq 1}^{N} f_{jk} f_{kl} f_{lj} P_{jkl}, \quad P_{jkl} = P_{jk} P_{kl}, \qquad (4.56)$$

the operators $\{P_{jk}\}$ are operators of permutation of any subjects with indices j, k. The function $F_{jkl} = f_{jk} f_{kl} f_{lj}$ and the operator P_{jkl} are both invariant under cyclic permutations of the idices (jkl). There are *two* invariants due to the dependence of f_{jk} on the spectral parameter α (4.52). Later, the whole tower of invariants of higher-order degrees of $\{P_{jk}\}$ was constructed [100], but their mutual commutativity has not been proved till now, being an open and rather complicated mathematical problem.

After this long introduction, it is possible to formulate the problem under consideration: independently, Sutherland *et al.* [99] constructed from the L matrix of the Lax pair the class of "dynamical spin" models with the Hamiltonian

$$\mathcal{H} = \frac{1}{2} \left[-\sum_{j=1}^{N} \left(\frac{\partial}{\partial x_j} \right)^2 + \sum_{j \neq k}^{N} a(a + P_{jk}) h_{jk} \right], \quad a \in \mathbf{R}, \qquad (4.57)$$

with $h_{jk} = \coth^2(x_j - x_k)$, i.e. for hyperbolic degeneration of the general elliptic case, integrability of which is still under question since they did not prove the commutativity of the integrals of motion which they found. As in the case of spin chains (4.55), Lax relation exists but does not help, and now one must also be guided by intuition and experience given in treatment of elliptic spin chains — in absence of any regular way of constructing integrals of motion. So, let us choose now $h_{jk} = -f_{jk} f_{kj} = \wp(x_j - x_k) - \wp(\alpha)$ in (4.57) (note that both periods of $\wp$ are now arbitrary). As before, the term with spectral parameter does not influence the dynamics: the operator $\sum_{J \neq k}^{N} P_{jk}$ trivially commutes with all Hamiltonians of the type (4.57) and might be omitted. Let us make an appropriate Ansatz

$$I = \sum_{j \neq k \neq l}^{N} \frac{\partial^3}{\partial x_j \partial x_k \partial x_l} + \sum_{j \neq k \neq l}^{N} \left[(F_{jk} + G_{jk} P_{jk}) \frac{\partial}{\partial x_l} \right.$$

$$\left. + \mu f_{jk} f_{kl} f_{lj} P_{jkl} \right], \tag{4.58}$$

where $P_{jkl} = P_{jk} P_{kl}$ is symmetric with respect to cyclic permutations of the indices (jkl) (as multiplication law of the elements of S_N dictates), F_{jk}, G_{jk} are some functions and μ is a factor. All undefined terms in (4.58) (F_{jk}, G_{jk} and μ) might be determined from the relation

$$[H, I] = 0 \tag{4.59}$$

if the Ansatz is correct. Let us consider I as a sum of three terms,

$$I_1 = \sum_{j \neq k \neq l}^{N} \frac{\partial^3}{\partial x_j \partial x_k \partial x_l},$$

$$I_2 = \sum_{j \neq k \neq 1}^{N} (F_{jk} + G_{jk} P_{jk}) \frac{\partial}{\partial x_l},$$

$$I_3 = \sum_{j \neq k \neq l}^{N} \mu f_{jk} f_{kl} f_{lj}$$

and write down the explicit calculation of $[H, I_1]$,

$$[H, I_1] = \left[\sum_{m \neq n}^{N} a \left(a + P_{mn} h_{mn}, \sum_{j \neq k \neq l}^{N} \frac{\partial^3}{\partial x_j \partial x_k \partial x_l} \right) \right].$$

Suppose first that $m = j, n = k$. Then

$$\sum_{l \neq m, n}^{N} \frac{\partial}{\partial x_l} \left[\sum_{m \neq n \neq l}^{N} a(a + P_{mn}) h_{mn}, \frac{\partial^2}{\partial x_m \partial x_n} \right]$$

$$= a^2 \sum_{l \neq m \neq n}^{N} \frac{\partial}{\partial x_l} \left[h_{mn}, \frac{\partial^2}{\partial x_m \partial x_n} \right].$$

Now suppose that $m = j$, $n \neq k, l$. Then

$$\left[\sum_{m \neq n}^{N} a(a + P_{mn})h_{mn}, \quad \sum_{m \neq n \neq l, n \neq k} \frac{\partial^3}{\partial x_m \partial x_k \partial x_l} \right]$$

$$= \sum_{m \neq k \neq l \neq n} a(a + P_{mn})(f'_{mn}f_{nm} - f_{mm}f'_{nm})\frac{\partial^2}{\partial x_k \partial x_l}.$$

After some long but not too tedious calculations, by using several times the functional equation (4.51), well-known "additional law" for Weierstrass functions

$$(\zeta_{jk} + \zeta_{kl} + \zeta_{lj})^2 = \wp_{jk} + \wp_{kl} + \wp_{lj}, \tag{4.60}$$

and symmetry of the operator $P_{jkls} = P_{jk}P_{kl}P_{ls}$ with respect to cyclic permutations of indices $(jkls)$, one finds that (4.59) is satisfied if and only if

$$F_{jk} = 3a^2 h_{jk}, \quad G_{jk} = 3a h_{jk}, \quad \text{and} \quad \mu = 2a^2 \tag{4.61}$$

for arbitrary values of spectral parameter α in (4.53). It means that there are *two* integrals of motion given by (4.58), as in the case of quantum spin chains with elliptic exchange (4.55). Indeed, $h_{jk} = \wp(x_j - x_k) - \wp(\alpha)$ by definition, and the function

$$f_{jk}f_{kl}f_{lj} = \frac{\sigma(x_j - x_k + \alpha)\sigma(x_k - x_l + \alpha)\sigma(x_l - x_j + \alpha)}{\sigma(x_j - x_k)\sigma(x_k - x_l)\sigma(x_l - x_j)\sigma^3(\alpha)} \tag{4.62}$$

is, as one can easily see, the elliptic function of α with only one third-order pole at $\alpha = 0$ on the torus $\mathbf{T} = \mathbf{C}/\mathbf{Z}\omega_1 + \mathbf{Z}\omega_2$ (note that, contrary to the case of spin chains (4.55), both ω_1 and ω_2 are now arbitrary). By calculation of first four terms in the Laurent decomposition of (4.25) as function of α near $\alpha = 0$, one finds with the use of (4.62) that

$$f_{jk}f_{kl}f_{lj} = -\frac{1}{2}\wp'(\alpha) + \wp(\alpha)[\zeta(x_j - x_k) + \zeta(x_k - x_l) + \zeta(x_l - x_j)]$$

$$- \frac{1}{6}(2[\zeta(x_j - x_k) + \zeta(x_k - x_l) + \zeta(x_l - x_j)]^3$$

$$+ \wp'(x_j - x_k) + \wp'(x_k - x_l) + \wp'(x_l - x_j)). \tag{4.63}$$

The first term in (4.63) gives nothing since the operator $\sum_{j\neq k\neq l}^{N} P_{jkl}$ trivially commutes with all Hamiltonians of the type (4.57). Two next terms are quite non-trivial and the Ansatz (4.58) with (4.61) gives *two* integrals of motion:

$$I_1 = -3a \sum_{j\neq k\neq l}^{N} P_{jk}\frac{\partial}{\partial x_l} + 2a^2 \sum_{j\neq k\neq l}^{N}$$

$$\times [\zeta(x_j - x_k) + \zeta(x_k - x_l) + \zeta(x_l - x_j)]P_{jkl},$$

$$I_2 = \sum_{j\neq k\neq l} \left(\frac{\partial^3}{\partial x_j \partial x_k \partial x_l} + 3a(a + P_{jk})\wp(x_j - z_k)\frac{\partial}{\partial x_l} \right)$$

$$- \frac{a^3}{3} \sum_{j\neq k\neq l}^{N} \{2[\zeta(x_j - x_k) + \zeta(x_k - x_l) + \zeta(x_1 - x_j)]^3$$

$$+ \wp'(x_j - x_k) + \wp'(x_k - x_l) + \wp'(x_l - x_j)\}P_{jkl}.$$

Note that I_1 is linear in momenta of particles. This fact is rather unexpected, and it is confirmed by independent calculation which shows that the relation $[H, I_1] = 0$ really takes place. It is much more easier than in general case (4.58) due to absence of third-order derivatives in I_1. The main step in the proof consists in using the formula

$$\wp'(y) - \wp'(x) = 2[\wp(y) - \wp(x)][\zeta(x + y) - \zeta(x) - \zeta(y)], \quad (4.64)$$

which can be easily obtained by comparing residues of Laurent decomposition of left- and right-hand sides of (4.64) as elliptic functions of arguments x and y.

It is natural to consider the question of mutual commutativity of the operators I_1, I_2. By very long calculations, it is found that these operators commute for all values of a. It seems that for $N = 3$ there are no other operators commuting with H.

There are some intriguing questions which might be of interest for further investigations. One of them is finding the way of constructing integrals of motion with higher degrees of derivatives and P-operators. It seems that there are many of them as in the case of elliptic spin chains. It would be highly desirable to simplify the proof of their mutual commutativity. And the last but not least consists in finding explicit formulae for $a \in \mathbf{N}$ for spin representation of S_N by Pauli matrices. Quite probably, they might be expressed via 1D Weierstrass sigma functions as it takes place for scalar case [101].

In the next section, it will be shown that all that takes place for the simplest non-trivial case of $SU(2)$ spins and lowest value of the integer parameter a.

4.4. Solution of Three-Particle Spin Elliptic Calogero–Moser System

In this section, the case of elliptic interaction between particles with spin $1/2$ will be considered in detail. At $N = 2$, $P_{jk} = 1$, $h_{jk} = \wp(x_j - x_k)$, Equation (4.20) reduces to usual Lamé equation, but the matrix problem is highly non-trivial even in the case $N \geq 3$, $n = 2$ (particles carry spin $1/2$) will be mainly discussed. The first results for spinless particles were obtained in [102–104], but only in [101] the explicit form of ψ were found for $a = 1$, $N = 3$ and algebraic structure of the manifold containing all ψ-s was described for all $a \in \mathbf{N}$. Later on, overcomplicated meromorphic solutions for all $a \in \mathbf{N}$ were found in [80]; in [81], the explicit formulae for all $a \in \mathbf{R}$ were obtained for spinless case in the form of infinite series.

The existence of quantum Lax relation $[H, L] = [L, M]$ was also mentioned in [81] for elliptic case, but it does not give the integrals of motion in the form $Ij = \sum_{k,l}^{N}(L^j)_{kl}$ since the M matrix does not obey the "sum-to-zero" conditions $\sum_{j=1}^{N} M_{jk} = \sum_{k=1}^{N} M_{jk} = 0$.

In this situation, every analytical result for the solutions of elliptic matrix Schrödinger equation is of value, even for some restriction for a, N and n. In what follows, we put $N = 3$ (three-particle case) and consider at first the question of integrability of the problem defined by (4.57) for arbitrary a and N. For the spinless case, it is known that there is the operator

$$
\mathcal{J}_{scalar} = \frac{\partial^3}{\partial x_1 \partial x_2 \partial x_3} + a(a+1) \left[(\wp(x_{23}) - \wp(\alpha)) \frac{\partial}{\partial x_1} \right.
$$

$$
\left. + (\wp(x_{31}) - \wp(\alpha)) \frac{\partial}{\partial x_2} + (\wp(x_{12}) - \wp(\alpha)) \frac{\partial}{\partial x_3} \right]
$$

which commutes with the Hamiltonian. For the spin case but in the absence of dynamical degrees of freedom, the additional integral of motion is also known as follows:

$$\mathcal{J}_{spin} = \sum_{j \neq k \neq l \neq j} f_{jk} f_{kl} f_{lj} P_{jk} P_{kl},$$

where $f_{jk} = f(n_j - n_k)$, $\{n_j\}$ are the positions of spins. Combining these expressions, one finds that the proper Ansatz for the integral of motion for particles with spin is

$$\mathcal{J} = \frac{\partial^3}{\partial x_1 \partial x_2 \partial x_3} + \frac{1}{2} \sum_{j \neq k \neq l \neq j}^{3} a(a + P_{jk})(\wp(x_{jk} - \wp(\alpha)) \frac{\partial}{\partial x_l}$$

$$+ \lambda \sum_{j \neq k \neq l \neq j}^{3} f(x_{jk}) f(x_{kl}) f(x_{lj}) P_{jk} P_{kl},$$

where the function $f(x_{jk})$ is the same as in the previous section,

$$f(x_{jk}) = \frac{\sigma(x_{jk} + \alpha)}{\sigma(x_{jk})\sigma(\alpha)},$$

and $\sigma(x)$ is the Weierstrass sigma function; λ is some parameter which should be determined by the commutativity condition,

$$[H, \mathcal{J}] = 0.$$

Direct computation shows that this Ansatz is indeed correct and

$$\lambda = \frac{a^2}{3}.$$

Due to the presence of spectral parameter α in $\mathcal{J}_{scalar}$ and in function $f(x_{jk})$, there are in fact two integrals of motion due to arbitrariness of this parameter,

$$\mathcal{J}_1 = \frac{\partial^3}{\partial x_1 \partial x_2 \partial x_3} + \sum_{j \neq k \neq l \neq j}^{3} \left(\frac{1}{2} a(a + P_{jk})\wp(x_{jk}) \frac{\partial}{\partial x_l} + \frac{a^2}{3} \varphi_{jkl} P_{jk} P_{kl} \right),$$

$$\mathcal{J}_2 = \sum_{j \neq k \neq l \neq j} \left(P_{jk} \frac{\partial}{\partial x_l} - \frac{a}{3} \psi_{jkl} P_{jk} P_{kl} \right),$$

where

$$\varphi_{jkl} = -\frac{1}{2}[\wp'(x_{jk}) + \wp'(x_{kl}) + \wp'(x_{lj}) + 2(\zeta(x_{jk} + \zeta(x_{kl}) + \zeta(x_{lj}))^3],$$

$$\psi_{jkl} = \zeta(x_{jk}) + \zeta(x_{kl}) + \zeta(x_{lj}),$$

$\zeta(x_{jk})$ is the Weierstrass zeta function. All these formulas were checked by the Mathematica program. The formula for $\mathcal{J}_2$ is especially simple: it resembles the total momentum (and coincides with it as $\{P_{jk}\} = 1$). In the trigonometric limit, it can be expressed through scalar product of the Yangian operator and total momentum, which (in this limit only!) both commute with the Hamiltonian. One confirms also, by direct computation of $[\mathcal{J}_1, \mathcal{J}_2]$, that these operators mutually commute and form with H and total momentum the commutative ring for all values of the parameter a.

Let us now construct the explicit solutions of the eigenvalue problem for the simplest non-trivial case of three particles carrying spin $1/2$ [120]. When all spins aligned up or down, one has the situation analogous to the spinless case. The non-trivial form of the wave function ψ arising for the states with total spin $1/2$ is as follows:

$$\psi(x_1, x_2, x_3) = A(x)|\uparrow\uparrow\downarrow\rangle + B(x)|\uparrow\downarrow\uparrow\rangle + C(x)|\downarrow\uparrow\uparrow\rangle,$$

$$A + B + C = 0. \tag{4.65}$$

The last equation means orthogonality to the total $S = 3/2$ subspace. The operators $a(a + P_{jk})$ act on the spin pairs in the states $(\uparrow\uparrow), (\downarrow\downarrow)$ as $a(a + 1)$ and for $(\uparrow\downarrow - \downarrow\uparrow)$ as $a(a - 1)$. If a is chosen as positive integer, there are singularities in the spinless case in the form of poles, $(x_j - x_k)^{-a}$ as $x_{jk} \to 0$. It is natural to expect that in the case of particles with spin at least some solutions to the eigenequation have the similar structure, i.e. A, B, C have singular behavior as $x_{jk} \to 0$ in the form of poles. In the simplest non-trivial case of $a = 1$, the eigen equatuion reads in the component form after substituting (4.65) as

$$\left(\frac{1}{2}\sum_{j=1}^{3}\frac{\partial^2}{\partial x_j^2} + E\right)A - \sum_{j>k}^{3}\wp(x_{jk})A$$

$$- [\wp(x_{12})A + \wp(x_{31})C + \wp(x_{23})B] = 0, \tag{4.66}$$

$$\left(\frac{1}{2}\sum_{j=1}^{3}\frac{\partial^2}{\partial x_j^2} + E\right)B - \sum_{j>k}^{3}\wp(x_{jk})B$$

$$- [\wp(x_{12})C + \wp(x_{31})B + \wp(x_{23})A] = 0, \qquad (4.67)$$

$$\left(\frac{1}{2}\sum_{j=1}^{3}\frac{\partial^2}{\partial x_j^2} + E\right)C - \sum_{j>k}^{3}\wp(x_{jk})C$$

$$- [\wp(x_{12})B + \wp(x_{31})A + \wp(x_{23})C] = 0. \qquad (4.68)$$

Let us introduce the notation

$$Y(x) = A(x) - B(x), \quad Z(x) = A(x) - C(x)$$

and deduct (4.67) and (4.68) from (4.66). Under the condition (4.65), it is easy to see that the system (4.66)–(4.68) is equivalent to two coupled equations for $Y(x)$ and $Z(x)$,

$$\left(\frac{1}{2}\sum_{j=1}^{3}\frac{\partial^2}{\partial x_j^2} + E\right)Y - \wp(x_{12})(Y + Z) - (2Y - Z)\wp(x_{31}) = 0,$$

$$(4.69)$$

$$\left(\frac{1}{2}\sum_{j=1}^{3}\frac{\partial^2}{\partial x_j^2} + E\right)Z - \wp(x_{12})(Y + Z) - (2Z - Y)\wp(x_{23}) = 0.$$

$$(4.70)$$

Since the "potentials" here are double periodic, it might be expected that the solutions to (4.69) and (4.70) are quasiperiodic, acquiring the same Bloch factors under the shifts of the arguments by the periods ω_1, ω_2 of the Weierstrass functions. According to (4.69), $Y(x)$ has a simple pole at $x_{31} \to 0$, the same for $Z(x)$ as $x_{23} \to 0$. The analysis of limits $x_{23} \to 0$ for (4.69) and $x_{31} \to 0$ for (4.70) shows that the left-hand sides of (4.69) and (4.70) are regular at these conditions. And finally, if $Y(x) \to Z(x)$ as $x_{12} \to 0$, there should be a simple pole singularity of these functions in this limit. Combining all these properties, one comes to the Ansatz for Y and

Z in the form

$$Y(x) = b\frac{\sigma(\mu_{12})\sigma(x_{12} + \lambda_{12})\sigma(x_{31} + \lambda_{31})}{\sigma(x_{12})\sigma(x_{31})}$$
$$\times \exp(k_1 x_1 + k_2 x_2 + k_3 x_3), \tag{4.71}$$

$$Z(x) = b\frac{\sigma(\lambda_{12})\sigma(x_{12} + \mu_{12})\sigma(x_{23} + \mu_{23})}{\sigma(x_{12})\sigma(x_{23})}$$
$$\times \exp(k_1 x_1 + k_2 x_2 + k_3 x_3), \tag{4.72}$$

where $\sigma(x)$ is the Weierstrass sigma function defined above; $b, \{k_j\}, \lambda_{12}, \lambda_{31}, \mu_{12}, \mu_{23}$ are some parameters. The Bloch factors for (4.71), (4.72) are equal if and only if

$$\mu_{12} = \lambda_{12} - \lambda_{31}, \quad \mu_{23} = -\lambda_{23}. \tag{4.73}$$

These expressions look rather asymmetric in $\{A, B, C\}$, but the symmetry becomes evident with the use of remarkable identity

$$Y(x) - Z(x) = C(x) - B(x)$$
$$= -b\frac{\sigma(\lambda_{31})\sigma(x_{23} + \lambda_{12})\sigma(x_{31} - \lambda_{12} + \lambda_{31})}{\sigma(x_{23})\sigma(x_{31})}$$
$$\times \exp(k_1 x_1 + k_2 x_2 + k_3 x_3),$$

which is valid for all values of the parameters λ_{12} and λ_{31} and coordinates $\{x_j\}$. The substitution of (4.71) and (4.72) into (4.69), (4.70) by direct but rather long calculations shows that (4.71) and (4.72) under the conditions (4.73) indeed give the solutions to the system (4.69), (4.70) if the following restrictions to the parameters $\{\lambda\}, \{k\}$ take place:

$$k_1 - k_2 = \zeta(\lambda_{31} - \lambda_{12}) - \zeta(\lambda_{12}), \tag{4.74}$$

$$k_2 - k_3 = \zeta(\lambda_{31}) + \zeta(\lambda_{12}), \tag{4.75}$$

where ζ is the Weierstrass zeta function. To get (4.74), (4.75), the formula

$$\zeta(x) + \zeta(y) + \zeta(z) - \zeta(x + y + z) = \frac{\sigma(x + y)\sigma(x + z)\sigma(y + z)}{\sigma(x)\sigma(y)\sigma(z)\sigma(x + y + z)}$$

was used. These formulas, (4.71) and (4.72), together with (4.74), (4.75) look very simple, being valid for $a = 1$ only. One can only guess what happens in the case $a > 1$; to our best knowledge, nobody has found the wave functions even for three-particle systems with spin despite there being a rather long but explicit expression for the case of spinless particles.

References

[1] M.A. Olshanetsky and A.M. Perelomov, *Phys. Rep.* 71, 314 (1981).

[2] M. Adler, *Commun. Math. Phys.* 55, 195 (1977).

[3] E.T. Whittaker, *Analytical Dynamics*, Vol. 332 (London, CUP, 1927).

[4] C. Holt, *J. Math. Phys.* 23, 1037 (1982).

[5] M.A. Olshanetsky and A.M. Perelomov, *Lett. Math. Phys.* 1, 187 (1976).

[6] V.I. Inozemtsev, *J. Phys.* A 17, 815 (1984).

[7] V.I. Inozemtsev, *Phys. Lett.* 98A, 316 (1983).

[8] A.M. Perelomov, *Lett. Math. Phys.* 1, 531 (1977).

[9] V.I. Inozemtsev, *Phys. Scripta* 29, 518 (1984).

[10] V.I. Inozemtsev and D.V. Meshcheryakov, *Phys. Lett.* 106A, 105 (1984).

[11] V.I. Inozemtsev, *Phys. Scripta* 39, 289 (1989).

[12] F. Calogero, *Lett. Nuovo Cim.* 19, 505 (1977).

[13] S. Ahmed, *Lett. Nuovo Cim.* 26, 285 (1979).

[14] A.P. Veselov, *Uspekhi Mat. Nauk* 35, 195 (1980) (in Russian).

[15] V.I. Inozemtsev, *Funct. Anal. Appl.* 23, 323 (1990).

[16] B.A. Dubrovin, *Uspekhi Mat. Nauk* 36, 11 (1981) (in Russian).

[17] B.A. Dubrovin, *Funct. Analiz i Prilozhen.* 11(4), 28 (1977) (in Russian).

[18] V.I. Inozemtsev and D.V. Meshcheryakov, *Lett. Math. Phys.* 9, 13 (1985).

[19] M. Toda, *J. Phys. Soc. Jpn.* 23, 501 (1987).

[20] H. Flashka, *I. Phys. Rev.* B9, 1924 (1974); *II. Progr. Theor. Phys.* 51, 703 (1974).

[21] J. Moser, *Batelle Recontres-Lecture Notes in Physics* 38, 468 (1974).

[22] O.I. Bogoyavlensky, *Commun. Math. Phys.* 51, 205 (1976).

[23] M.A. Olshanetsky and A.M. Perelomov, *Invent. Math.* 54, 261 (1979).

[24] A.G. Reyman and M.A. Semenov-Tyan-Shansky, *Invent. Math.* 54, 81 (1979).

[25] A.G. Reyman and M.A. Semenov-Tyan-Shansky, *Invent. Math.* 63, 423 (1981).

[26] M.A. Olshanetsky, A.M. Perelomov, A.G. Reyman, and M.A. Semenov-Tyan-Shansky, and *Contemporary Problems of Mathematics* 16, 86 (1987).

[27] M. Adler, and P. van Moerbeke, *Adv. Math.* 38, 267 (1980).

[28] M. Adler and P. van Moerbeke, *Adv. Math.* 38, 318 (1980).

[29] H. Ochiai, T. Oshima and H. Sekiguchi, *Proc Japan Acad.* 70, Ser. A, 62 (1994).

[30] V.I. Inozemtsev and D.V. Meshcheryakov, *Phys. Lett.* 108A, 315 (1985).

[31] R. Sasaki and K. Takasaki, *J. Phys. A, Math. Gen.* 34, 9533 (2001).

[32] S. Ahmed, *Lett. Nuovo Cimento* 26, 285 (1979).

[33] I.M. Krichever, *Funct. Anal. Appl.* 14, 282 (1980).

[34] J. Dittrich and V.I. Inozemtsev, *J. Phys. A, Math. Gen.* L999 (1993).

[35] O.V. Chalykh and A.P. Veselov, *Commun. Math. Phys.* 126, 597 (1990).

[36] P. Appell, *Bull. Soc. Math. France* 10, 59 (1882).

[37] G. Felder and A. Varchenko, *Int. Math. Res. Notices* 5, 222 (1985).

[38] V.I. Inozemtsev and D.V. Meshcheryakov, *JINR Rapid Communications*, No.4, 22 (1984).

[39] V.I. Inozemtsev and D.V. Meshcheryakov. *Phys. Lett.* 111A,234 (1985).

[40] V.I. Inozemtsev and D.V. Meshcheryakov, *Phys. Scripta* 23, 99 (1986).

[41] V.l.S. Dotsenko and V.A. Fateev, *Phys. Scripta* 21, 112 (1984).

[42] W. Heisenberg, *Z. Phys.* 49, 619 (1928).

[43] P.A.M. Dirac, *Proc. Roy. Soc.* 123A, 714 (1929).

[44] J.H. van Vleck, *The Theory of Electric and Magnetic Susceptibilities* (Oxford, Clarendon Press, 1932).

[45] H.A. Bethe, *Z. Phys.* 71, 205 (1931).

[46] M. Gaudin, *La fonction d'onde de Bethe* (Paris, Masson 1983).

[47] L.D. Faddeev, in *Recent Advances in Field Theory and Statistical Mechanics.* (Amsterdam: North-Holland, 1984), p. 561.

[48] V.E. Korepin, N.M. Bogoliubov and A.G. Izergin, *Quantum Inverse Scattering Method and Correlation Functions* Cambridge, (Cambridge University Press, 1993).

[49] F. Calogero, *J. Math. Phys.* 12, 419 (1971).

[50] B. Sutherland, *J. Math. Phys.* 12, 246, 251 (1971); *Phys. Rev.* A5, 1372 (1972).

[51] F. Calogero, G. Marchioro and O. Ragnisco, *Lett. Nuovo Cim.* 13, 383 (1975).

[52] F. Calogero, *Lett. Nuovo Cim.* 13, 411 (1975); *J. Moser. Adv. Math.* 16, 197 (1975).

[53] I.M. Krichever, *Funct. Anal. Appl.* 14, 282 (1980).

[54] M.A. Olshanetsky and A.M. Perelomov, *Phys. Rep.* 94, 313 (1983).

[55] F.D.M. Haldane, *Phys. Rev. Lett.* 60, 635 (1988); 66, 1529 (1991).

[56] B.S. Shastry, *Phys. Rev. Lett.* 60, 639 (1988).

[57] F.D.M. Haldane, A. Okiji and N. Kawakami (eds.), In *Proceedings of the 16th Taniguchi Symposium on Condensed Matter Physics*, cond-mat/9401001.

[58] V.I. Inozemtsev, *J. Stat. Phys.* 59, 1143 (1990).

[59] V.I. Inozemtsev and N.G. Inozemtseva, *J. Phys.* A24, L859 (1991).

[60] J. Sekiguchi, *Publ. RIMS Kyoto Univ.* 12, 455 (1977).

[61] O.A. Chalykh and A.P. Veselov, *Commun. Math. Phys.* 126, 597 (1990).

[62] V.I. Inozemtsev, *Commun. Math. Phys.* 148, 359 (1992).

[63] V.I. Inozemtsev, *Lett. Math. Phys.* 28, 281 (1993).

[64] V.I. Inozemtsev and B.-D. Dörfel, *J. Phys.* A26, L999 (1993).

[65] J. Dittrich and V.I. Inozemtsev, *J. Phys.* A26, L753 (1993).

[66] E.K. Sklyanin, *Progr. Theor. Phys. Suppl.* 118, 35 (1995).

[67] H. Frahm and V.I. Inozemtsev. *J. Phys.* A27, L801 (1994).

[68] A.P. Polychronakos, *Phys. Rev. Lett.* 70, 2329 (1993); H. Frahm, *J. Phys.* A26, L473 (1993).

[69] F. Calogero, *Lett. Nuovo Cim.* 19, 505 (1977).

[70] V.I. Inozemtsev, *Phys. Scripta* 39, 289 (1989).

[71] V.I. Inozemtsev, *J. Phys.* A28, L439 (1995).

[72] V.I. Inozemtsev, *J. Math. Phys.* 37, 147 (1996).

[73] V.I. Inozemtsev, *Lett. Math. Phys.* 36, 55 (1996).

[74] K. Hasegawa, *Commun. Math. Phys.* 187, 289 (1997).

[75] J. Dittrich and V.I. Inozemtsev, *Mod. Phys. Lett.* B11, 453 (1997).

[76] V.I. Inozemtsev, *Phys. Lett.* 98A, 316 (1983).

[77] V.I. Inozemtsev and N.G. Inozemtseva, *J. Phys.* A30, L137 (1997).

[78] M. Takahashi, *J. Phys.* C10, 1289 (1977).

[79] J. Dittrich and V.I. Inozemtsev, *J. Phys.* A30, L623 (1997).

[80] G. Felder and A. Varchenko, *Int. Math. Res. Notices* 5, 222 (1995).

[81] V.I. Inozemtsev, *Regul. Chaotic Dyn.* 5, 236 (2000).

[82] B. Sutherland and B.S. Shastry, *Phys. Rev. Lett.* 71, 5 (1993).

[83] J. Hubbard, *Proc. Roy. Soc. London* A276, 238 (1963).

[84] E.H. Lieb and F.Y. Wu, *Phys. Rev. Lett.* 20, 1445 (1968).

[85] B.S. Shastry, *J. Stat. Phys.* 50, 57 (1988).

[86] F. Gebhard and A.E. Ruckenstein, *Phys. Rev. Lett.* 68, 214 (1992).

[87] F. Gebhard, A. Girndt and A.E. Ruckenstein, *Phys. Rev.* B49, 10926 (1994).

[88] P.A. Bares and F. Gebhard, *Europhys. Lett.* 29, 573 (1995).

[89] F.D.M. Haldane *et al.*, *Phys. Rev. Lett.* 69, 2021 (1992).

[90] V.G. Drinfeld, *Soviet. Math. Dokl.* 32, 254 (1985).

[91] F. Göhmann and V.I. Inozemtsev, *Phys. Lett.* A214, 161 (1996).

[92] D.B. Uglov and V.E. Korepin, *Phys. Lett.* A190, 238 (1994).

[93] V.I. Inozemtsev and R. Sasaki, *Phys. Lett.* A289 (2001).

[94] F. Gebhard, *The Mott Metal-Insulator Transition.* Springer, Heidelberg, 1997.

[95] Z.N.C. Ha and F.D.M. Haldane, *Phys. Rev.* B46, 9359 (1992).

[96] F. Calogero, *Lett. Nuovo Cimento* 13, 411 (1975).

[97] M.A. Olshanetsky and A.M. Perelomov, *Phys. Rep.* 94, 313 (1983).

[98] V.I. Inozemtsev, *J. Stat. Phys.* 59, 1146 (1990).

[99] B. Sutherland and B.S. Shastry, *Phys. Rev. Lett.* 71, 5 (1993); I. Cherednik, *Adv. Math.* 106, 65 (1996).

[100] V.I. Inozemtsev, *Lett. Math. Phys.* 36, 55 (1996).

[101] J. Dittrich and V.I. Inozemtsev, *J. Phys.* A26, L753 (1993).

[102] E.H. Lieb and W. Liniger, *Phys. Rev.* 130, 1605 (1963).

[103] J.B. McGuire, *J. Math. Phys.* 5, 622 (1964); 7, 123 (1966).

[104] F. Calogero, *J. Math. Phys.* 2191, 2197 (1969); 12, 419 (1971).

[105] B. Sutherland, *J. Math. Phys.* 12, 247, 251 (1971).

[106] B. Sutherland, *Phys. Rev.* A5, 1372 (1972).

[107] V.I. Inozemtsev, *Phys. Scripta* 29, 518 (1984).

[108] V.I. Inozemtsev and D.V. Meshcheryakov, *Phys. Scripta* 33, 99 (1986).

[109] N. Kawakami, *Phys. Rev.* B46, 3191 (1992).

[110] J.A. Minahan *et al.*, *Phys. Lett.* B302, 265 (1993).

[111] D. Bernard *et al.*, *J. Phys. A: Math. Gen.* 26, 5219 (1993).

[112] K. Hikami and M. Wadati, *J. Phys. Soc. Jpn.* 59, 4203 (1993).

[113] K. Vacek, A. Okiji and N. Kawakami, *Phys. Rev.* B49, 4635 (1994).

[114] K. Vacek, A. Okiji and N. Kawakami, *J. Phys.* A27, L201 (1994).

[115] Y. Kato and Y. Kuramoto, *Phys. Rev. Lett.* 74, 1222 (1995).

[116] K. Hikami, *Nucl. Phys.* B441, 530 (1995).

[117] H. Frahm and V.I. Inozemtsev, *J. Phys.* A27, L801 (1994).

[118] T. Yamamoto, *J. Phys. Soc. Jpn.* 63, 1212 (1994).

[119] L. Brink *et al.*, *Phys. Lett.* 286B, 109 (1996).

[120] J.C. Barba and V.I. Inozemtsev, *Phys. Lett.* A372, 5951 (2008).

Printed in the United States
by Baker & Taylor Publisher Services